AF608874

Optimization Methods for Integrating Energy and Production Systems

Optimierungsmethoden für die Integration von Energie- und Produktionssystemen

Von der Fakultät für Maschinenwesen der
Rheinisch-Westfälischen Technischen Hochschule Aachen
zur Erlangung des akademischen Grades eines Doktors
der Ingenieurwissenschaften genehmigte Dissertation

vorgelegt von

Ludger Leenders geb. Holters

Berichter: Univ.-Prof. Dr.-Ing. André Bardow
Univ.-Prof. Alexander Mitsos, Ph.D.

Tag der mündlichen Prüfung: 14. April 2022

Aachener Beiträge zur Technischen Thermodynamik Band 35

Ludger Leenders
Optimization Methods for Integrating Energy and Production Systems
Optimierungsmethoden für die Integration von Energie- und Produktionssystemen

ISBN: 978-3-95886-445-0

Bibliografische Information der Deutschen Bibliothek
Die Deutsche Bibliothek verzeichnet diese Publikation in der Deutschen Nationalbibliografie; detaillierte bibliografische Daten sind im Internet über http://dnb.ddb.de abrufbar.

Herstellung & Vertrieb:

1. Auflage 2022
© Wissenschaftsverlag Mainz GmbH - Aachen
Süsterfeldstr. 83, 52072 Aachen
Tel. 0241 / 87 34 34 00
www.Verlag-Mainz.de

ISSN: 2198-4832

Satz: nach Druckvorlage des Autors
Umschlaggestaltung: Druckerei Mainz

printed in Germany
D82 (Diss. RWTH Aachen University, 2022)

Danksagung

Die vorliegende Arbeit entstand am Lehrstuhl für Technische Thermodynamik der RWTH Aachen University sowie am Energy & Process Systems Engineering Lab der ETH Zürich im Rahmen meiner Tätigkeit als wissenschaftlicher Mitarbeiter. Mein besonderer Dank gilt meinem Doktorvater Prof. Dr.-Ing. André Bardow. Er hat mir stets die Freiheit gegeben die Promotion in allen Facetten selbst zu gestalten, hat mich bei allen Vorhaben unterstützt und mir das nötige Vertrauen entgegengebracht. Ich danke außerdem für die Gelegenheit, mit ihm an die ETH Zürich zu wechseln und die Anfänge seiner neuen Forschungsgruppe begleiten zu können, was mich fachlich und persönlich geprägt hat. Mein Dank gilt auch Prof. Alexander Mitsos, Ph.D., für die Übernahme des Korreferates und ganz besonders für die fruchtbare Zusammenarbeit im Bereich der Bileveloptimierung. Außerdem möchte ich mich bei Herrn Prof. Dr.-Ing. Jakob Andert für die Übernahme des Prüfungsvorsitzes bedanken.

Ich bedanke mich bei allen Kolleg:innen des LTT und EPSE, die durch ihren großartigen Teamgeist die Zeit der Promotion unvergesslich gemacht haben. Ein großer Dank geht an die Energiesystemtechnikgruppe am LTT und die Systemgruppe am EPSE. Im Besonderen danke ich dem Optimierungsteam und dem E$T Sailing Team, mit denen ich nicht nur eine gute Zusammenarbeit sondern auch Freundschaft und einen unvergesslichen Segeltrip verbinde. Zusätzlich möchte ich mich bei meiner ehemaligen Gruppenleiterin Dr.-Ing. Maike Hennen und bei meinem ehemaligen Gruppenleiter Dr.-Ing. Matthias Lampe bedanken, die stets ein offenes Ohr für alle Belange hatten. Auch danke ich Gregor Migas, ohne dessen Support ich viel mehr Zeit mit IT Angelegenheiten verbracht hätte. Ein Dank geht auch an die Projektpartner des Forschungsprojektes *imPROvE*, in dessen Rahmen diese Arbeit entstanden ist. Dabei danke ich vor allem Dr.-Ing. Franz Lanzerath von TLK Energy für die hervorrangende Leitung des Projektes. Die Zusammenarbeit im Projekt war stets fachlich interessant und hat mir viel Freude bereitet.

Allen meinen Bürokolleg:innen danke ich für die tolle Atmosphäre und die schönen Erinnerungen, wie dem Büroausflug nach Zingst. Dabei möchte ich mich besonders bei Dr.-Ing. Björn Bahl bedanken für mehr als drei Jahre hervorragende Betreuung seit meiner Masterarbeit. Ich möchte mich aber nicht nur für die gemeinsame Zeit im Büro bedanken, sondern auch für die schönen Stunden außerhalb des Büros. Andreas Kämper danke ich insbesondere für die Unterstützung und Geduld am Ende der Promotion, das gute Büroklima, die gemeinsamen Espressi und viele fachliche und besonders gute persönliche Diskussionen. Außerdem bedanke ich mich bei Niklas Nolzen für die Unternehmungen, die wir in Zürich außerhalb unseres Büroalltags unternehmen und die schöne Freundschaft, die sich daraus entwickelt hat. Dr.-Ing. Sarah von Pfingsten danke ich für die kurze, aber tolle gemeinsame Zeit in 101.2. Bedanken möchte ich mich auch bei Dr.-Ing. Nils Klinkenberg für die geteilte Teamleitung, die mir viel Freude bereitet hat und bei der wir uns super ergänzt haben. David Shu möchte ich für den schönen gemeinsamen Start am EPSE in Zürich danken und für die gute Zeit abseits des Promotionsalltags. Bedanken möchte ich mich auch bei Dr.-Ing. Hatim Djelassi für die hervorragende Zusammenarbeit zum Thema Bileveloptimierung, bei der ich viel von ihm lernen konnte. Zudem danke ich allen Student:innen, vor allem Dörthe Hagedorn, Alissa Ganter, Kirstin Ganz, Thomas Kohne, Anna Starosta und Stefan Rauscher, für ihren großen Beitrag zu meiner Promotion und die großartige Zusammenarbeit, in der ich viel gelernt habe und Freude hatte.

Zu guter Letzt gilt mein Dank den vielen Personen außerhalb des Arbeitsumfeldes, die für mich immer ein wertvoller Ausgleich zur Promotion waren: Allen Kletterfreunden, dem Stammtisch, der ESG Big Band und dem AV. Ein besonderer Dank geht dabei an Lukas ter Huurne, mit dem mich seit über drei Jahrzehnten die beste Freundschaft verbindet. Ganz außerordentlich möchte ich meinen Eltern, Geschwistern und meiner Familie danken. Meinen Eltern gilt dabei ein besonderer Dank dafür, dass sie mich in allen Vorhaben stets unterstützt haben. Der größte Dank geht an meine Frau Theresa, für ihre Unterstützung bei und neben der Promotion, der ich mir immer sicher sein konnte. Mit ihrer Leichtigkeit und Begeisterungsfähigkeit wirkt jede Herausforderung direkt viel kleiner. Ich freue mich auf alles, was das Leben noch mit uns gemeinsam vorhat!

Zürich, im Mai 2022 *Ludger Leenders*

'A problem is a chance for you to do your best.'

Duke Ellington

Contents

Appendices

List of Figures

List of Tables

Notation

Abbreviations and acronyms

AC	absorption chiller
B	boiler
CC	compression chiller
c.f.	confer (English: compare)
CHP	combined-heat-and-power engine
E	production equipment
e.g.	exempli gratia (English: for example)
El	electricity-driven
GCG	General-Column-Generation
i.e.	id est (English: that is)
inc.	incomplete
inf.	information
LBD	lower bound
LLP	lower-level problem
LP	linear programming
MILP	mixed-integer linear programming
MILBP	mixed-integer linear bilevel programming
MINLP	mixed-integer non-linear programming
NEG	negative control reserve
NO	no control reserve
POS	positive control reserve
PV	photovoltaic
S	storage unit and state
STN	State-Task-Network
T	task
UBD	upper-bounding problem

Latin symbols

a	time step and parameter
A	matrix of parameters
b	parameter and vector of parameters
B	batch size
c	specific energy cost
C	cost
CM	maintenance cost
E	energy
f	objective function
M	large number
mf	maintenance factor
n	number
o	binary for operational state
OC	operational costs
p	price
P	power and profit
$product$	produced material
pvaf	present value annuity factor
q	interest rate
R	revenue
raw	raw material
t	time
u	relative input
U	input power
v	relative output
V	output power and capacity
w	upper-level variables
W	binary for batch start
x	continuous variable
y	binary for installed energy conversion unit

Greek symbols

α	machine environment and binary for selected energy price band
β	characteristics and binary for valid discretization point
γ	scheduling objective and binary for existence
Δ	difference
ϵ	small value and optimality tolerance
ζ	binary for control-reserve scenario
η	efficiency
θ	binary for operational limit
κ	binary for selected linear section
λ	binary for selected energy price
ξ	time-dependent capacity
ω	probability

Subscripts and superscripts

aux	auxiliary
B	boiler
bilvl	bilevel
buy	buying
CAP	capacity
CAPEX	capital expenditure
CHP	combined-heat-and-power engine
con. frac.	consumption fraction
corr. int.	corrected integrated
cost	optimized cost
curr	current
cycle	periodic cycle
D	discretized
demand	energy demand
Discr. Point	discretization point
el	electricity
ES	energy system
est	estimated cost
fix	fixed

free	free energy conversion units
gas	natural gas
grid	electricity grid
heat	heating
i	number of iteration
inc. inf.	incomplete information
int	integrated
k	energy price band
l	lower level
lb	lower price band and lower bound
LBD	lower bound
LLP	lower-level problem
lower	small decrease and lower operational limit
max	maximal
min	minimal
N	nominal
NEG	negative control reserve
NO	no control reserve
obj	objective
off	idle units
offer	offered cost
on	running units
OPEX	operational expenditure
piping	piping unit
POS	positive control reserve
process	processing
prod	produced
prod. frac.	production fraction
PS	production system
R	control reserve
RP	control-reserve provision
sell	selling
seq	sequential
supply	energy supply
TAC	total annualized cost
tot	total revenue
TOTAL	total
u	upper level

ub	upper price band and upper bound
UBD	upper-bounding problem
upper	large increase and upper operational limit
var	variable
year	year
zero	idle energy conversion units

Sets

$cr \in CR$	control reserve
$cu \in CU$	electricity consuming units
$d \in D$	discretized and other energy conversion units
$e \in E$	energy form
$ep \in EP$	energy price
$f \in F$	number of free energy conversion units
$g \in G$	linear sections for operation
$h \in H$	linear sections for capacity
$i \in I$	tasks
$j \in J$	production equipment
$k \in K$	discretization points
M	generic set
$pu \in PU$	electricity producing units
$s \in S$	storage units
$t \in T$	time
$u \in U$	energy conversion units
$(u,e_1,e_2) \in C$	energy forms supplied by energy conversion units

Kurzfassung

Die wichtigste Maßnahme zur Verminderung des Klimawandels ist die Reduzierung der Treibhausgasemissionen. Eine Schlüsselrolle hierbei spielt die energieintensive Industrie, da der industrielle Sektor große Mengen an Treibhausgasen emittiert. Ein Großteil dieser Treibhausgasemissionen wird durch den Energieverbrauch verursacht. Daher ist eine effizientere Energieversorgung der Industrie notwendig.

An grossen industriellen Standorten versorgen oft dezentrale Energiesysteme die Produktionssysteme. Beide Systeme optimieren dabei den Betrieb hinsichtlich einer Zielfunktion wie beispielsweise Betriebskosten oder Gewinn. Die vorliegende Arbeit präsentiert Optimierungsmethoden für solche industrielle Standorte. Die Optimierungsmethoden berücksichtigen dabei unterschiedliche Beziehungen zwischen den Systemen, wobei zwei Fälle unterschieden werden: Im ersten Fall verfolgen die Systeme die gleiche Zielfunktion, im zweiten Fall individuelle Zielfunktionen. Eine gleiche Zielfunktionen besteht beispielsweise, wenn beide Systeme einem Unternehmen angehören. Individuelle Zielfunktionen bestehen, wenn beide Systeme beispielsweise unterschiedlichen Unternehmen angehören.

Für den Fall, dass Energie- und Produktionssystem das gleiche Ziel verfolgen, wird eine Methode für das integrierte Design beider Systeme präsentiert und eine Methode für die integrierte Bereitstellung von Regelenergie. Für den Fall, dass Energie- und Produktionssystem individuelle Zielfunktionen verfolgen, werden wiederum zwei Fälle unterschieden: unvollständiger und vollständiger Informationsaustausch. Für den unvollständigen Informationsaustausch wird eine Optimierungsmethode für die Koordination zwischen einem Energie- und einem Produktionssystem vorgestellt. Diese Optimierungsmethode wird anschließend für den Fall mehrerer Energie- und Produktionssysteme weiterentwickelt. Liegt vollständiger Informationsaustausch zwischen den Systemen vor, wird ein Bilevelproblem formuliert. Für die Lösung des Bilevelproblems wird ein existierender Lösungsalgorithmus angepasst.

Die in dieser Arbeit entwickelten Methoden werden auf Fallstudien angewandt und Vor- und Nachteile untersucht. Die Fallstudien zeigen, dass keine Methode in allen identifizierten Beziehungen zwischen den Systemen die für das Produktionssystem optimale Lösung ermittelt. Daher ist je nach vorliegendem Fall die jeweilige Optimierungsmethode anzuwenden. Zusammenfassend stellt diese Arbeit Optimierungsmethoden für alle identifizierten Beziehungen zwischen Energie- und Produktionssystemen bereit. Dadurch ermöglicht diese Arbeit die Auswahl einer geeigneten Optimierungsmethode für alle Arten von Produktionssystemen mit dezentraler Energieversorgung.

Abstract

The key measure to mitigate climate change is the reduction of greenhouse gas emissions. Hereby, energy-intensive industry plays a key role due to its substantial greenhouse gas emissions. A substantial share of these greenhouse gas emissions is caused by energy supply. Thus, energy supply needs to be more efficient in industry.

In large industrial sites, on-site energy systems often supply production systems. Both systems thereby optimize their operation with respect to an objective such as operational cost or revenue. This thesis provides optimization methods for these large industrial sites. The optimization methods reflect two relationships between both systems: Both systems can either follow the same objective or system-specific objectives. The same objective exists, e.g., if both systems belong to one company. System-specific objectives exist, e.g., if both systems belong to different companies.

For the case that both systems follow the same objective, a method is presented for the integrated synthesis of both systems. For the same case, a method is presented for integrated scheduling to provide control reserve. For the case that energy and production systems have system-specific objectives, two cases are distinguished: incomplete and complete information exchange. For incomplete information exchange, an optimization method is introduced for the coordination between a single energy and a single production system. This optimization method is then extended to multiple energy and multiple production systems. For complete information exchange between the systems, a bilevel problem is formulated. For solving the bilevel problem, an existing solution algorithm is adapted.

All methods presented in this thesis are applied to case studies, and advantages and disadvantages are examined. The case studies show that no method provides the optimal solution for the production system in all identified relationships between the systems. Thus, depending on the case at hand, the respective optimization method has to be applied. Overall, this thesis presents optimization methods for all identified relationships between energy and production systems. Thus, this thesis enables the selection of a suitable optimization method for all kind of production systems with decentralized energy supply.

Chapter 1

Introduction

The industrial sector is the economic sector emitting the largest share of greenhouse gases, being responsible for more than 30 % of greenhouse gas emissions. Around 80 % of industrial greenhouse gas emissions are energy-related emissions (IPCC, 2014). Thus, to reach climate goals, improving the energy supply of the industrial sector is highly important.

Energy is supplied to industrial production systems via grids, transport or on-site energy systems. These energy systems need be decarbonized. Decarbonization of energy systems can be reached by replacing the fossil basis of energy-conversion technologies with renewables. Another option for decarbonization is to increase the efficiency of energy supply and use. Thereby, the energy system can become more efficient even without any investment.

The efficiency of energy supply and use by on-site energy and production systems can be improved by integrating the systems operation. Production schedules impact not only production cost but also the resulting energy demand. The energy demand impacts the operation of the energy system, and thus, the efficiency of energy supply. The integrated scheduling of both systems improves the efficiency by exploiting synergies.

In practice, the optimizations of energy and production systems are commonly performed separately. First, the production system is scheduled, and subsequently, the energy system schedules for the resulting energy demand. To integrate both optimizations, research on energy and production systems needs to be combined. For example, chemical production systems are often supplied by on-site energy systems. Thus, on the one hand, Energy Systems Engineering needs to be extended to consider the energy consumer. On the other hand, scheduling of chemical production systems is a major area in Process Systems Engineering and should be extended by the energy supply. Researchers in Process Systems Engineering already identified the need for extension (Pistikopoulos et al., 2021).

When merging Energy and Process System Engineering, new challenges arise. One challenge is to properly consider whether the energy and production systems follow the same or misaligned objectives. If the objectives are the same, an integrated optimization of energy and production system can be performed. For this purpose, interconnections between the energy and production system need to be modeled. If the objectives of the individual systems are misaligned, e.g., two different companies operate the systems, an optimization problem with an overall objective cannot be applied, because each system would follow its own objective. Individual systems with misaligned objectives can often be described as a Stackelberg game. In a Stackelberg game, one system is the leader and the other system is the follower. This relationship can be further be distinguished depending on whether the systems have complete or incomplete information on each other. In this work, we[1] propose methods to optimize production and on-site energy systems in all the identified relationships with the same or misaligned objectives.

1.1 Structure of this thesis

In Chapter 2, first, the state-of-the-art of optimizing energy and production systems is reviewed. Afterward, we state the contribution of this thesis.

The following Chapters 3-7 are structured in two parts: In Part I, we present methods for the integrated optimization of energy and production systems following the same objective. In Part II, we present methods to optimize energy and production systems in a Stackelberg game, and thus, following misaligned objectives.

Part I: Integrated optimization

In Chapter 3, we present a method for the integrated synthesis of energy and batch production systems. As decision variables, the design, size and operation of the production and the energy system are optimized. For this purpose, we combine an existing MILP optimization model of a batch production system with an existing MILP optimization model of an energy system. The chapter shows how the two individual models are combined to an integrated optimization problem.

In Chapter 4, we use the MILP optimization model from Chapter 3 and fix the design and size of energy and batch production system. We extend the resulting scheduling problem to include the participation in control-reserve markets. The pro-

[1]This thesis is written in the *pluralis modestiae* to avoid the excessive use of passive voice. Furthermore, I emphasize that the research behind this thesis is done in collaboration with colleagues, supervised students and my supervisor. The contribution of the author is pointed out in the beginning of each chapter.

posed method identifies a production schedule, which leads to an energy demand that allows the energy system to increase profits from control-reserve provision.

Part II: Stackelberg-based optimization

In Chapter 5, we present a method to schedule energy and production systems when the systems are in a Stackelberg game and exchange incomplete information. We introduce the exchange of incomplete information between the systems. The exchanged incomplete information is only based on a few energy demands and their corresponding cost for supply. The proposed so-called demand-dependent energy costs are sufficient to realize substantial savings.

In Chapter 6, we extend the method from the previous chapter from single to multiple energy and production systems. The resulting challenge is that we have to coordinate between production systems and energy systems as well as among the production systems and among the energy systems. For this purpose, we propose coordination methods.

In Chapter 7, we formulate the bilevel problem to schedule energy and production systems when the systems have misaligned objectives but exchange complete information. We solve the bilevel problem by specializing the algorithm from Djelassi et al. (2019) that solves the bilevel problem to a predefined gap and identifies the best solution for the Stackelberg game.

In Chapter 8, this thesis is summarized and conclusions are drawn. Finally, the author discusses future perspectives of scheduling integrated energy and production systems and identifies future research topics.

Chapter 2

State-of-the-art in optimization of energy and production systems

In Section 2.1, a review is given for scheduling and synthesis of energy and production systems. In Section 2.2, we review optimization methods for integrated systems. In Section 2.3, we outline the contribution of this thesis.

Leenders, L., Bahl, B., Hennen, M., and Bardow, A. (2019a). Coordinating scheduling of production and utility system using a Stackelberg game. *Energy*, 175, 1283-1295.

Leenders, L., Bahl, B., Lampe, M., Hennen, M., and Bardow, A. (2019b). Optimal design of integrated batch production and utility systems. *Computers & Chemical Engineering*, 128, 496-511.

Leenders, L., Starosta, A., Baumgärtner, N., and Bardow, A. (2020). Integrated scheduling of batch production and utility systems for provision of control reserve. In *Proceedings of ECOS 2020 - 33rd International Conference on Efficiency, Cost, Optimization, Simulation and Environmental Impact of Energy Systems*, pages 712-723. Osaka, Japan.

Leenders, L., Ganz, K., Bahl, B., Hennen, M., Baumgärtner, N., and Bardow, A. (2021). Scheduling coordination of multiple production and utility systems in a multi-leader multi-follower Stackelberg game. *Computers & Chemical Engineering*, 150, 107321.

Leenders, L., Hagedorn, D. F., Djelassi, H., Bardow, A., and Mitsos, A. (2022). Bilevel optimization for joint scheduling of production and energy systems. *Optimization & Engineering*.

The author of this thesis contributed to the development and implementation of the methods as well as the calculation and interpretation of the results, wrote the draft of the papers and is their principal author.

2.1 Scheduling and synthesis of energy and production systems

In this section, we start with a general introduction to scheduling and continue with a review focusing on scheduling of production systems and energy systems.

2.1.1 Scheduling of systems

The definition and characterization of scheduling in this section are mainly based on Pinedo (2016) and Maravelias (2012). We present some basic fundamentals. However, for a extensive review on these fundamentals, the reader is referred to the literature, e.g., Maravelias (2012); Pinedo (2016).

Pinedo (2016) defines scheduling as a decision-making process. Commonly, scheduling is performed repetitively to allocate *resources* to *tasks*, where one or multiple *objectives* are optimized (Pinedo, 2016). Scheduling is applied in all kinds of industries to optimize the schedule of a system, e.g.:

- scheduling of production systems (Kopanos and Puigjaner, 2019)
- scheduling of energy systems (Mancarella, 2014)
- production planning in supply chains (Voss and Woodruff, 2006)
- planning of human resources in healthcare systems (Hall, 2012)
- resource scheduling in cloud computing (Zhan et al., 2015)

The system can deliver a physical product or a service.

The tasks in a scheduling problem can be very different, e.g., a manufacturing stage in a car factory, printing a book in a print shop or work packages in an engineering project. Resources can be human resources, machines, trains, production sites, etc.

The resources to be scheduled mainly operate in batch operation or continuous operation. In batch operation, a task is performed for a single charge. All desired material is assembled to the resource before the task starts. An example is a reaction in a vessel where the vessel is filled, and the reaction takes place over a few hours. Once the task is complete, the product is withdrawn.

In continuous operation, a task is performed continuously, and material is assembled non-stop. An example is the operation of a heat pump where the electricity is supplied by the grid and heat is continuously produced. There are also other operation modes which are mostly a variation of batch and continuous operation, e.g., in semi-batch operation, some material is assembled while the task is performed.

Scheduling problems can be described by the triplet $\alpha|\beta|\gamma$ (Pinedo, 2016): α describes the environment. β describes the characteristics of the system and can contain multiple entries. γ describes the scheduling objective. The environment in the α field describes the topology and how the tasks have to be processed through the system. In production scheduling, the environment is often named the machine environment, and the machines are the resources. Machines can operate in parallel if they process the same task. These parallel machines are then often named as a *step*. As we focus on energy and production systems, we will use these terms in the following.

The operation of the machines is scheduled such that the desired *jobs* are finished. A job is an order that needs to be satisfied by the system and typically consists of a predefined set of tasks. For this purpose, the job has to visit the steps that perform the needed tasks.

Scheduling has intensively been studied in the chemical industry and is a major area in Process Systems Engineering. The scheduling problems considered in this thesis belong to the research field of Process Systems Engineering. In Process Systems Engineering, the machine environment can be more complex, as described by Maravelias (2012): In discrete manufacturing, a job corresponds to a specific batch that is maintained through the production process. The identity of the batch is not changed during its processing through the machine environment. In contrast, in chemical production environments, batches can often be mixed or split. Therefore, chemical production environments are classified by the handling of the material instead of the structure of the machines. For a more detailed comparison of discrete manufacturing and chemical production, the reader is referred to Maravelias (2012).

In the β field, the system to be scheduled is further characterized. A system needs to be described by many characteristics, e.g., release dates, no-wait constraints, processing time of a batch etc. The γ field describes the objective function of the scheduling problem which is maximized or minimized. The objective functions can be both single objective or multi objective (Maravelias, 2012; Pinedo, 2016).

The scheduler has to identify and consider the relevant properties of the system, i.e., the entries of the triplet $\alpha|\beta|\gamma$. Scheduling a system can become complex very quickly. Nowadays, scheduling problems are not solved by hand but instead are commonly modeled and solved by computer programs.

Modeling of scheduling problems

Once the system to be scheduled is identified, a model of the system has to build up. For this purpose, the modeler identifies the suitable model formulation. One aspect of the model formulation used for scheduling is the time representation (Floudas and Lin, 2004; Méndez et al., 2006; Pinedo, 2016): discrete time or continuous time. In discrete-time representation, a predefined time grid is defined, and events can only happen at the beginning of a time period in this time grid. Often, the time grid consists of time steps with an equal length. Thus, many time points might be necessary to take the different lengths of tasks into account. In continuous-time representation, events can take place at any time. Commonly, fewer time points are needed. For the representation of events, different concepts are available: global time intervals, unit-specific time events, global time points, etc. Discrete-time models only use global time intervals. In comparison, different event presentations are possible for continuous-time models such as global time points or unit-specific time events (Méndez et al., 2006).

If the needed time representation is chosen, other aspects can be added to the model. For example, the modeler can consider uncertainties for the parameters, e.g., unknown product or energy demand, availability of resources. Uncertainties in models can be considered by different approaches: stochastic programming (Birge and Louveaux, 2011; Klein Haneveld et al., 2020) or robust optimization (Soyster, 1973; Bertsimas and Sim, 2004). In stochastic optimization, the probability distribution of the uncertain parameters is considered in the decision-making process. In contrast, in robust optimization, the objective of the worst-case scenario is optimized. Both optimization techniques are active research fields.

In the synthesis of a system, scheduling is considered to anticipate the operation costs. When synthesizing a system, the system is often scheduled for a representative period, e.g., day or week. The identification of representative periods is a challenging task (Bahl et al., 2018; Baumgärtner et al., 2019b; Teichgraeber and Brandt, 2019). The optimal synthesis of systems is a wide and active research field. Here, we interpret the synthesis of a system as using a superstructure and deciding whether a machine is built. Furthermore, we determine the size of the machine. Still, the author is aware that this is a very simplified view, and this view neglects many challenges.

Scheduling and synthesis problems are often formulated as mathematical programs involving integer decisions. Thus, many scheduling problems are mixed-integer linear programs or mixed-integer nonlinear programs. A wide variety of methods are available for the solution, e.g., branch-and-bound algorithms, evolutionary algorithms, dynamic programming (Pinedo, 2016). In recent years, machine learning attracts much attention and is also applied to solve scheduling problems, e.g., by using reinforcement learning (Hubbs et al., 2020).

This thesis focuses on optimization methods for scheduling integrated energy and production systems which are therefore discussed next.

2.1.2 Scheduling and synthesis of energy systems

In the last decades, major advances have been achieved in scheduling and synthesis of energy systems (Mancarella, 2014; Andiappan, 2017; Frangopoulos, 2018; Demirhan et al., 2019). In particular, Demirhan et al. (2019) present possible future research directions such as increased interdisciplinary work. The increase of interdisciplinary work is one focus of this thesis.

In the following, we provide a review focusing on scheduling energy systems in general and scheduling energy systems to provide control reserve. Also, a short review on the synthesis of energy systems is provided.

2.1.2.1 Scheduling of energy system

The IPCC (2014) defines an energy system consisting of different components. The components produce, convert, deliver or use energy. An energy system can be of different scales. For example, an energy system can be the air-conditioning system of a car. An energy system can also take the size of a whole country or even multiple interconnected countries. The scale of the energy system that is considered depends on the questions to be answered.

In this thesis, we focus on energy systems supplying industrial sites. These on-site energy systems typically operate several energy conversion units. Furthermore, we focus on multi-energy systems which supply more than one energy form at a time. The multi-energy character results in complex interactions between the energy conversion units (Mancarella, 2014).

Scheduling of energy systems identifies for each time step which energy conversion unit supplies energy such that the overall energy demand is fulfilled. An objective is minimized, e.g., cost, greenhouse-gas emissions.

Energy-system models often consider time-dependent input data, for example, time-dependent electricity prices. The consideration of uncertainty in the electricity prices complicates the model. The uncertainties often result from the uncertain energy production of renewable energy technologies. Renewable energy can be highly fluctuating and difficult to predict. Thus, forecasting methods are needed for, e.g., electricity generation from wind (Zhang et al., 2014) and photovoltaic (Sobri et al., 2018).

Uncertainties are not only relevant for the operation of renewable energy technologies within on-site energy systems. The uncertainties from renewable generation are also challenging for the electricity grid that supplies the on-site energy system with electricity since supply and demand need to be balanced at all times.

Control-reserve provision

The electricity grid balances unplanned fluctuations in supply and demand by control reserve. Control reserve is a balancing service offered by electricity providers or consumers. The providers or consumers offer to increase or decrease their electricity production or consumption depending on grid requirements, respectively.

Decentralized energy systems can supply control reserve. Decentralized energy systems have different opportunities to supply the control reserve, such as overcapacities, flexible units, and storage units. To identify the amount of control reserve offered, scheduling models are utilized. Muche et al. (2016) model the participation of a biomass-fueled combined-heat-and-power plant in the positive tertiary reserve market. If control-reserve energy is requested, the authors ensure feasibility by storing excess heat produced by the combined-heat-and-power plant to be used another day. Kumbartzky et al. (2017) model a combined-heat-and-power plant with a flexible power-to-heat ratio to participate in the tertiary reserve market and the day-ahead spot market. The uncertain prices in both markets are considered in a multi-stage stochastic programming model. Scenarios are modeled for the reserve capacity price and the spot market price. No energy price is considered for the requested control reserve. Zhang et al. (2021) schedule a virtual power plant that provides control reserve under various uncertainties. Uncertainties are considered for the energy prices, wind power availability, and the request of control reserve.

The literature shows that methods are available to consider the provision of control reserve in scheduling of energy systems. However, control reserve is commonly not supplied by utilizing the multi-energy connection between the energy conversion units in an energy system. For example, if an increase in electricity supply is requested, a combined-heat-and-power engine can increase its electricity and, in consequence, heat production. An absorption chiller can use the excess heat to supply cooling. To meet the cooling demand, a running compression-chiller can reduce its supplied cooling, and thus, the electricity consumption which increases the net electricity supply. Thereby, the multi-energy connection between the energy conversion units is exploited. Utilizing the multi-energy interconnection within the energy system would be beneficial.

Additionally, the literature above shows that control-reserve provision commonly focuses on the integration of the decentralized energy system with the grid. However, the energy system simultaneously interacts with the energy consumer who could support the provision of control reserve. A consideration of the energy system and the energy consumer for control-reserve provision seems desirable. Furthermore, the consideration of the energy consumer, and thus, the demand side of the energy system seems desirable in general.

Demand-side management

Demand-side management summarizes all activities on the energy system's demand side (Palensky and Dietrich, 2011). Demand-side management is, e.g., performed when a large energy consumer such as an aluminum plant changes its electricity demand as well as when the local energy consumer of the decentralized energy system changes its energy demand. A change in the local energy demand can result in a more efficient schedule of the decentralized energy system. In addition, the energy demand of the energy system can be changed without changing the energy demand of the energy consumer. An example is the reduction of thermal energy demands by heat integration and setting up a heat exchanger network (Papoulias and Grossmann, 1983b; Klemeš and Kravanja, 2013). Also, without changing the energy demand of the energy consumer, energy storage can be used to shift the energy demand of the energy system in time.

Demand-side management is also performed when the energy demand of the energy consumer is changed. Still, the potential of an energy demand change by the energy consumer can be identified only by analyzing the energy demand. For this purpose, to identify demand-side management potential for energy systems, Bahl et al. (2017) provide a method independently from a specific system of the energy consumer. In their method, the benefits from reductions in the energy demand are identified without any model of the energy consumer.

Zhang and Grossmann (2016) discuss demand-side management for enterprise-wide optimization and identify the integration of energy and production management as promising. The integration could adapt the energy demand and thus the production schedule to enable a more efficient operation of the energy systems.

Concluding this review on energy system scheduling, energy systems can be improved by demand-side management. The demand side could be managed such that the scheduling of the energy system is integrated with the scheduling of the energy consumer, e.g., production system. Furthermore, energy systems face the challenge of uncertainties such as production from renewable energies and the provision of control reserve. However, most energy systems have flexibility in their operation, which can be utilized by considering the uncertainties in scheduling.

2.1.2.2 Synthesis of energy systems

In the synthesis of an energy system, the components of the system and their interconnections are designed. Synthesis of energy systems is pioneered by Papoulias and Grossmann (1983a). Their model for the synthesis of energy systems considers a constant energy demand. In general, energy demands are not constant. Thus, energy systems need to be synthesized for time-dependent energy demands.

Today, most approaches synthesize energy systems for time-dependent energy demands (Andiappan, 2017) or even for time-dependent grid emissions (Baumgärtner et al., 2019a). Also the synthesis of energy systems can take into account uncertainties either by stochastic optimization (Zhou et al., 2013; Mavromatidis et al., 2018; Nolzen et al., 2021) or robust optimization (Majewski et al., 2017a,b). Andiappan (2017) reviews optimization methods to handle uncertainties and multiple objectives in energy system synthesis. Furthermore, specific areas are discussed, such as retrofit, reliability, and flexibility. As research directions, Andiappan (2017) identifies to focus on reliability aspects rigorously, considering failure in the synthesis of energy systems, etc. Thus, many open research questions exists in energy system synthesis.

Typically, synthesis problems of energy systems are nonlinear and involve discrete decisions. The nonlinearities are caused by performance curves and investment cost curves of the energy conversion units. Discrete decisions involve the existence of an energy conversion unit and its on/off operation. Thus, synthesis problems of energy systems result in mixed-integer nonlinear programming problems (MINLP) (Goderbauer et al., 2016). However, the nonlinearities are often linearized to result in a mixed-integer linear programming problem (MILP) (Voll et al., 2013; Kämper et al., 2021a,b). This thesis employs MILP models for energy system synthesis and scheduling, since MILP models are often solved faster then the MINLP models while good solutions are found (Kämper et al., 2021a).

In energy system synthesis, commonly, the energy demand is considered as being already fixed. Similar to scheduling energy systems, an integrated view on the energy consumer and the energy system could be beneficial during the synthesis. Thereby, an optimal overall system can be synthesized.

In the following, we review scheduling and synthesis of production systems and, afterward, present methods to optimize integrated energy and production systems.

2.1.3 Scheduling and synthesis of production systems

In the last decades, major advances have been achieved in scheduling of production systems, and the topic is still an active research field (Maravelias, 2012; Harjunkoski et al., 2014; Castro et al., 2018; Letsios et al., 2020). The research focuses on many aspects, e.g., integration of decision making on different levels such as integration of scheduling and control (Dias and Ierapetritou, 2016) and integration of scheduling and planning (Dias and Ierapetritou, 2017), and energy-related aspects (Gahm et al., 2016).

In the following, we review production system scheduling. In the review, we focus on production scheduling and the consideration of energy consumption as the aspect most relevant for this thesis. For this purpose, we first review methods for production scheduling with a focus on energy-related scheduling. Afterward, we present a short review of synthesis methods for production systems.

2.1.3.1 Scheduling of production systems

Scheduling of production systems is a wide and active research field and important for almost all kinds of industries (Harjunkoski et al., 2014; Castro et al., 2018). In general, a large variety of production systems exists, resulting in many classes of scheduling problems. Operation types in production systems can be batch operation, continuous operation, or variations of batch and continuous (Section 2.1.1). In this thesis, we focus on production systems running in batch operation. However, in the review of methods considering energy aspects in scheduling, we also consider other operation types because the methods can sometimes be adapted for batch operation. Over the last decades, significant progress has been made to solve different classes of scheduling problems for production systems efficiently.

Very common formulations for scheduling problems are the State-Task-Network and the Resource-Task-Network formulations. The State-Task-Network (STN) formulation was proposed in Kondili et al. (1993), and Shah et al. (1993), the Resource-Task-Network (RTN) formulation was proposed by Pantelides (1994). As the names already suggest, the STN formulation employs states and tasks, where the RTN formulation employs resources and tasks. The RTN formulation handles identical equipment more efficiently compared to the STN formulation by a unified treatment of the resources.

However, when alternative equipment or processing times are modeled that depend on the machine, the tasks need to be duplicated in the RTN formulation (Méndez et al., 2006). Still, formulations for scheduling are developed to efficiently solve scheduling problems (Lee and Maravelias, 2017; Ackermann et al., 2021).

Energy in production system scheduling

Recently, research in production scheduling increasingly focuses on energy aspects (Gahm et al., 2016), where new challenges arise from more volatile electricity prices and more energy supply from renewable energy sources (Merkert et al., 2015; Mitsos et al., 2018). In batch scheduling, the energy demand was early considered by Kondili et al. (1993). The authors set resource constraints on the maximal availability of utilities during scheduling. Other energy-consumption-related issues are also considered in the literature, e.g., heat integration (Pinto, 2003; Seid and Majozi, 2015).

Recently, increasingly fluctuating electricity prices motivate to consider electricity consumption in the scheduling of production systems: Castro et al. (2011); Mitra et al. (2012); Zhang et al. (2016c); Basán et al. (2018) and Zhao et al. (2018) schedule production systems and take into account time-sensitive electricity prices. Hadera et al. (2019) decompose the production scheduling and electricity procurement and solve the problem by Mean Value Cross Decomposition. Mean Value Cross Decomposition (Holmberg, 1992) is a modification of Cross Decomposition (van Roy, 1983) which is a combination of the Benders decomposition and the Dantzig-Wolfe decomposition. Zhang et al. (2016a) schedule a production process and take into account uncertainties in the electricity price and the product demand. The uncertainties are considered within a two-stage stochastic program. Leo et al. (2021) also consider the uncertainty of the electricity price in day-ahead markets and deal with the uncertainty by a two-stage stochastic program. Thus, for production system scheduling, the consideration of uncertainties is still a field of interest in research (Grossmann et al., 2016). For example, uncertainty resulting from the unknown request of control reserve.

Control-reserve provision

The provision of control reserve is not only of interest for energy systems but also for production systems. In the previous review of energy system scheduling, we discussed that the request of control reserve is uncertain. Thus, the system that provides control reserve needs to be flexible such that the supply or demand of electricity from the grid can be adapted quickly. If a production system is flexible enough, it also can provide control reserve. The possibility of control-reserve request has to be modeled in the scheduling model of the production system. For this purpose, Zhang et al. (2015) schedule an air separation process with energy storage to participate in energy and control reserve markets. Robust optimization is utilized to consider the worst-case for control-reserve request. Zhang et al. (2016b) model a continuously operated production system and provide control reserve by interruptible load. The uncertainty of request is again considered by robust optimization. Schäfer et al. (2019) consider the participation in the control-reserve market for an energy-intensive process. The authors propose a decomposition method, where the two-stage problem is divided into two optimization problems. The two optimization problems are a nonlinear problem for the bidding strategy and a mixed-integer linear problem to schedule the production process. Bohlayer et al. (2020) schedule an energy-intensive production system that participates in energy and control-reserve markets. Their multi-stage stochastic program considers the stage-wise revelation of information in the different markets and provides the optimal bids for the markets.

The previously reviewed articles for control-reserve provision considered continuously operated production systems, which are assumed to interrupt their production process or transition into part-load operation. However, if a batch is started, the batch can often not be interrupted or operated in part-load. Thus, the production schedule is usually fixed and cannot be changed if control reserve is requested. Consequently, the reviewed methods might not be applicable for batch production systems. Still, if a batch production system is supplied by an on-site energy system, new opportunities arise to supply control reserve. Even if the production system schedule is fixed during operation, the ability of the energy system to supply control reserve can be increased by integrated optimization. For example, a production schedule can be identified that allows for a better control-reserve provision by the energy system. For this purpose, the production system should know the benefits resulting from a changed energy demand.

On-site energy supply

Many production scheduling problems consider only electricity purchase from the grid and neglect on-site electricity supply and other utilities. However, some authors consider the on-site supply of electricity, e.g., Hadera et al. (2015, 2019). On-site electricity supply is modeled by a constant tariff for electricity purchase with additional start-up costs. In practice, the costs for energy from on-site supply are not constant but depend on the operational state of the energy system.

The ability to consider true energy costs has a great potential for further cost reduction in scheduling production systems, as pointed out in the review by Harjunkoski et al. (2014). Thus, taking the operations of the on-site energy system into account may result in economic gains, and thus, is an excellent target for enterprise-wide optimization (Wassick, 2009). The challenge is to formulate the interconnection of energy and production systems. The interconnection unfixes the energy demand that as to be supplied by the energy system in the energy system optimization, and production systems consider that energy costs can change drastically depending on the energy demands. As a result, improved production schedules can be identified that lower cost from energy supply.

2.1.3.2 Synthesis of production systems

In the synthesis of a production system, the components of the system and their interconnections are designed. For the synthesis of continuously operated production systems, detailed models are used. For chemical production processes, even the molecules used within the production system are optimized in the synthesis (Papadopoulos et al., 2018). The synthesis of batch production systems often considers a less detailed view of the system, but takes the scheduling of the production system into account. The production schedule has a high impact on the synthesis problem due to the intermittent operation of the equipment (Barbosa-Póvoa, 2007). The configuration of each production step, e.g., a chemical reactor, can be optimized in more detailed optimizations prior to the synthesis of the batch production systems.

Reviews on the synthesis of batch production systems can be found in Barbosa-Póvoa (2007) and Majozi et al. (2015). However, the synthesis of batch production systems is often considered as an extension of scheduling and is a relatively small research field.

The synthesis problem of batch production systems can be further complicated by the consideration of energy. A challenging topic is the simultaneous synthesis of production systems and heat exchanger networks. Pinto (2003) combines the synthesis of batch production systems with heat integration to reduce energy consumption. The resulting design specifies the equipment, the operational schedule, and the areas of the corresponding heat exchangers. Seid and Majozi (2015) also present a model for the synthesis of batch production systems that considers heat exchangers. For their case study, the heat-integrated design increases the profit by 20 % and saves 40 % in energy requirements. Recently, Magege and Majozi (2021) propose a new formulation for the simultaneous synthesis, scheduling, and intermittent stream heat integration of production systems.

The reviewed studies take the synthesis of the heat integration into account during the synthesis of the batch production system. Still, methods are missing that consider the synthesis of the energy system simultaneously with the synthesis of the batch production system.

As for energy systems (c.f. Section 2.1.2.2), models for synthesis and scheduling production systems result in MINLP models due to nonlinear performance and investment cost curves. The nonlinearities can be linearized by piecewise-affine functions. In this thesis, we employ MILP models for production system scheduling like most scheduling problems (Castro et al., 2018).

So far, we have reviewed the scheduling and synthesis of energy and production systems. We identified that a combined optimization of both systems is favorable. In the following section, we provide an overview of optimization methods for integrated systems.

2.2 Optimization methods for integrated systems

On-site energy systems interface to different stakeholders, e.g., the production systems as customers and gas and electricity grids as their suppliers. When optimizing the operation of the on-site energy systems, these interfaces are typically represented by fixed input data for the energy system, i.e., energy demands and price data.

Often, each energy and production system on the industrial site is operated by a different company or business unit. As a result, the operation is optimized for each system individually (Engell et al., 2015). The individual optimizations are commonly performed sequentially: First, the production system schedules its production plan, which also defines its energy demand. Subsequently, the energy system individually schedules its operation to fulfill the energy demand. However, it seems highly beneficial to consider the interconnection between the systems during optimization.

If the systems are operated by one company and have a joint objective, an integrated optimization is suitable to improve the overall objective of the company. However, an integrated optimization might not be applicable if the systems have misaligned objectives or only share incomplete information about their objectives and constraints.

Game theory accounts for the rational decision-making of each participant in a game (Osborne and Rubinstein, 1994). A special game in game theory is the Stackelberg game (von Stackelberg, 1934). In a Stackelberg game, the systems are categorized as being either a leader or a follower. The leader decides first, and subsequently, the follower decides according to the leader's decision.

In industrial sites, the production system sets the energy demand first and can be considered as the leader. Subsequently, the energy system optimizes the operation to supply the energy demand of the production system and can be considered as the follower. Often, the systems in the industrial site are operated by different companies and do not reveal all information due to confidentiality. Thus, the systems do not reveal their objectives and constraints. Consequently, the production system has only incomplete information on the energy system. For the incomplete information exchange, methods for coordination can be applied. Most methods dealing with incomplete information iterate between the systems until a satisfying solution is reached. The other case is that the production system has complete information on the energy system. The Stackelberg game with complete information can be modeled as a bilevel problem (Dempe and Zemkoho, 2020). In a bilevel problem, the follower's optimization problem is embedded in the constraints of the leader. Thus, the leader takes into account that the follower has its own objective.

In the following sections, we review methods and applications for integrated optimization, for coordination between systems, and for bilevel optimization.

2.2.1 Integrated optimization

The integration of commonly separated operation problems is beneficial in general if both optimization problems share the same objective function. By the integration, synergistic effects can be employed.

For industrial sites, the integrated optimization of energy and production systems outperforms the sequential optimization of energy and production systems (Agha et al., 2010; Zhao et al., 2014; Zulkafli and Kopanos, 2016). The benefits of the integration are promising for both integrated scheduling and integrated synthesis.

In their pioneering work, Papoulias and Grossmann (1983c) investigate the integrated synthesis of production system, energy system, and heat recovery network. The investigated system operates continuously and in a single time step. Today, the integrated synthesis or scheduling of production systems and energy systems for continuous processes is still subject of current research.

Zhao et al. (2014) integrate the planning of multiple periods for the production system and the energy system. The proposed model saves substantially cost for a real-world example and also reduces the emissions of harmful gases. Zulkafli and Kopanos (2016) consider cleaning tasks as well as performance degradation and recovery for the energy system while they simultaneously schedule energy and production system. In their case study, they show that the savings rise when the prices for external energy purchases increase. The authors extend their work in Zulkafli and Kopanos (2017) by an improved model for performance degradation, modeling operational and maintenance policies, and considering startup and shutdown constraints. Pablos et al. (2021) perform an integrated scheduling of an industrial process and an on-site cogeneration unit. The authors consider the dominant process dynamics and the scheduling of the cogeneration unit. The method shows significant cost savings in demand response programs.

For integrated synthesis, Zhang et al. (2019) propose a multi-scale model of a network that produces electricity and renewables-based fuels. The model simultaneously designs the process units and the energy conversion units while taking operational constraints and time-varying availability of renewables into account.

The previously reviewed articles considered production systems running in continuous operation. To the best of our knowledge, Agha et al. (2010) present the first approach for integrated scheduling of batch production and on-site energy system. The authors demonstrate that the approach leads to significant savings in energy costs and greenhouse gas emissions compared to sequential scheduling. Furthermore, Agha et al. (2010) identify a higher potential of cost savings when the production system is not running at its full capacity and therefore allows more flexibility in the production schedule. The same authors proposed the Extended Resource Task Network (Théry et al., 2012), a formalism to schedule integrated energy and batch production systems. Subsequently, other authors looked into integrated energy and production system scheduling. Wang et al. (2012) propose an integrated scheduling problem of a batch production system and on-site energy unit to produce electricity. Hadera et al. (2016) use multi-parametric programming to solve the integrated scheduling of batch production scheduling and the electricity cost minimization. The electricity cost minimization considers several time-sensitive electricity contracts and onsite generation. The electricity cost minimization is solved separately using a multi-parametric program.

The above review shows that methods are available for the integrated scheduling of production systems and energy systems. However, methods are missing for the integrated scheduling for control-reserve provision by the interconnected systems. Additionally, methods are missing for the integrated synthesis of energy and batch production systems. For control-reserve provision, the methods need to consider that the production schedule is fixed for batch production systems and cannot be changed when control reserve is requested. Furthermore, the interconnection between the system needs to be considered such that synergistic effects can be employed.

While the integrated optimization of energy and production systems is beneficial for the overall system, energy and production systems are often operated by different internal business units of the same company or even by different companies. Thus, a single objective function can often not be defined in practice, and coordination methods or bilevel optimization problems need to be considered.

2.2.2 Coordination methods for optimization

Distributed optimization can be performed to consider that each system optimizes its operation individually and exchanges incomplete information (Wang et al., 2021). In distributed optimization, local optimization problems are solved, and the solution of the overall problem is identified by coordination between the individual optimization problems. The coordination of systems by distributed optimization is performed in many research areas, e.g., smart grids, multi-robot collaboration and transportation vehicle management. (Li et al., 2020). Distributed optimization can be performed if either an integrated optimization is computationally intractable or the systems only share incomplete information between each other.

Regarding the coordination between energy and production systems, there are parallels between smart grids and industrial sites: Smart grids need coordination between the energy supplier and consumer. In an industrial site, systems need to be coordinated if they share energy or other resources. Thus, in the following, we review methods for the coordination in smart grids and industrial sites. We focus on the exchanged information between the systems, which is a key factor regarding the performance of algorithms for distributed optimization.

For a smart grid, Atzeni et al. (2013) propose a distributed and iterative algorithm for demand-side management. In the algorithm, each user in the smart grid receives incomplete information from the other users in the form of aggregated energy loads of all users. Bahrami and Sheikhi (2016) propose an integrated demand response program between smart energy hubs producing heat and electricity, and utility companies. In an iterative approach, the energy price and demands are iteratively updated until a convergence criterion is met.

Coordination in industrial sites is relevant for the scheduling of energy and production systems. Thus, in the following, we review coordination methods applied to industrial sites. In the coordination of industrial sites, information on the resource prices is a key information to be exchanged between the systems. Furthermore, a central entity such as an auctioneer or coordinator manages the pricing and allocation of shared resources. Examples for exchanging the price information and using a central entity can be found in the following: Jose and Ungar (2000) coordinate in a chemical plant using an auctioneer. Cheng et al. (2007) optimize the operation in a plant-wide model predictive control problem using a coordinator. Wenzel et al. (2020) propose an algorithm to coordinate between different production plants in a large industrial site and exchange information via an aggregator and a coordinator. The algorithm in Wenzel et al. (2020) is based on many earlier works of the authors, e.g., in Wenzel et al. (2016), an algorithm is proposed to adjust the prices for shared resources. Allman and Zhang (2020) use the method of Alternating Direction Method of Multipliers for distributed optimization (Boyd et al., 2010) to coordinate the optimization between different companies, i.e., an industrial process and its customers. The industrial process as the central decision maker can also be interpreted as a coordinator.

In the reviewed articles, the price information on the resources is the main information to coordinate. However, the so far reviewed coordination methods are not explicitly developed for Stackelberg games and thus require a central entity for coordination. As introduced in Section 2.2, the situation between a production system and an energy system operated by different entities can be modeled as a Stackelberg game. In Stackelberg game with a leader and a follower, no central entity is necessary. Thus, in the following, we review methods that are developed for Stackelberg games using incomplete information.

For the coordination between energy suppliers and consumers, many approaches described the situation as a Stackelberg game. A single-leader multi-follower Stackelberg game is solved by Soliman and Leon-Garcia (2014),Yu and Hong (2016) and Motalleb et al. (2018). A multi-leader multi-follower Stackelberg game is solved by Maharjan et al. (2013, 2016), Wei et al. (2017), Mondal et al. (2018) and Alsalloum et al. (2020). We summarize the assignment of leader and follower and the exchanged incomplete information in Table 2.1. The reviewed methods for coordination between energy suppliers and consumers focus on exchanging energy demands and energy prices as incomplete information. In smart grids, the leader is commonly the energy supplier, and the energy consumer is the follower. The energy supplier is set as leader since it defines the energy price first for a large amount of energy consumers, and the energy consumers optimize their energy demand for the proposed prices. This is different in Mondal et al. (2018), where the energy consumers are the leader, because in the setup, first the energy consumer decide the request of energy. In industrial sites, the assignment of leader and follower is also reversed compared to the common setup in smart grids: The production system needs to run its production and identifies the energy demand first. Second, the energy system has to fulfill the energy demand and optimizes its operation to supply the energy.

In addition, there are no methods available to coordinate the operation of a batch production system with the operation of an energy system. For the coordination between these systems, the complex cost structure of the follower, i.e., energy system, must be captured by the incomplete information exchanged.

The cost structure of energy supply of an energy system is complex and the cost can change drastically with the energy demand. The complex cost structure is caused by nonlinear and discontinuous relationships between energy demand and energy price. In addition, the multi-energy character of the on-site energy system results in additional complexity that must be reflected by the exchanged incomplete information. Thus, only providing the energy price for a single energy demand might be insufficient information for the production system to anticipate the energy system behavior. However, while the cost structure must be captured by the incomplete information exchanged, the information still has to respect confidential requirements. Therefore, as we reviewed previously, energy costs and demand is often accessible information and should be used.

2.2.3 Bilevel optimization

In a Stackelberg game with complete information, the production system knows the optimization problem of the energy system. Thus, the production system can anticipate the cost of the energy system for a proposed energy demand exactly.

The energy cost of the production system depends directly on the optimization of the energy system. In the given situation of a Stackelberg game, the production system identifies the best solution, if it considers the energy systems optimization problem in the constraints. For this purpose, the production system can formulate and solve a bilevel problem.

Bilevel problems consider misaligned objective functions of an upper level (leader) and a lower level (follower) (Dempe and Zemkoho, 2020). Thus, bilevel problems model Stackelberg games where the leader decides first and the follower decides second depending on the decision of the leader. The sequential decision process is realized by embedding the lower-level optimization problem in the constraints of the upper-level optimization problem.

Table 2.1: Leader and follower assignment in the reviewed literature on Stackelberg games between energy supplier and energy consumer. Additional exchanged information is given after the reference. EP: energy price; ED: energy demand; ES: energy supply.

leader	exchanged information	follower	example
energy supplier	⟵ EP,ED ⟶	energy consumer	Maharjan et al. (2013) Soliman and Leon-Garcia (2014) Yu and Hong (2016) Maharjan et al. (2016) (+ES) Wei et al. (2017) Motalleb et al. (2018) (+other) Alsalloum et al. (2020)
energy consumer	⟵ EP,ED ⟶	energy supplier	Mondal et al. (2018)

Bilevel problems are challenging to solve and are proven to be NP-hard (Jeroslow, 1985; Bard, 1991). However, promising solution algorithms have been developed to solve bilevel problems. For solving bilevel problems with a nonconvex lower-level problem, solution algorithms are available for: nonlinear bilevel programming problems (NLBP) (Mitsos et al., 2008; Tsoukalas et al., 2009a,b; Wiesemann et al., 2013; Kleniati and Adjiman, 2014a,b), mixed-integer nonlinear bilevel programming problems (MINLBP) (Mitsos, 2010; Kleniati and Adjiman, 2015; Djelassi et al., 2019), and mixed-integer linear bilevel programming problems (MILBP) (Zeng and An, 2014; Hemmati and Smith, 2016; Fischetti et al., 2016; Yue et al., 2019). Thus, algorithms are available to solve bilevel problems.

Besides the development of general solution algorithms to solve bilevel problems, solution algorithms have been tailored to specific problems. Yue and You (2017) propose a solution algorithm for supply-chain optimization with complete information in a single-leader single-follower Stackelberg game. The algorithm solves the resulting MILBP problem. Ramos et al. (2018) proposed a single-leader multi-follower and a multi-leader single-follower Stackelberg game for the utility-network design of eco-industrial parks. Depending on the considered Stackelberg game, the authority of the eco-industrial park is the single leader/follower, and the enterprises are the multi follower/leader. In the eco-industrial park, the enterprises are continuously operating production plants. The bilevel problem is solved for the single-leader multi-follower Stackelberg game by replacing the convex follower problems with the Karush-Kuhn-Tucker reformulation. The bilevel problem for the multi-leader single-follower Stackelberg game is solved by a previously proposed method by the authors (Ramos et al.,

2016). Avraamidou and Pistikopoulos (2019) proposed an algorithm based on multi-parametric programming to solve a MILBP. In their case study, the algorithm designs and schedules a process with two production stages for three products. Yokoyama et al. (2019) solve a MILBP between a central power utility system and a distributed co-generation system. The problem is solved by an algorithm that uses a Karush-Kuhn-Tucker reformulation. Kostarelou and Kozanidis (2021) propose a heuristic solution algorithm to solve a MILBP for the optimal price-bidding in day-ahead electricity markets. In the MILBP, the upper level is the electricity producer, and the lower level is the electricity consumer. The heuristic solution algorithm is compared to an exact solution algorithm and shows fewer computational requirements.

Wogrin et al. (2020) review applications of bilevel optimization in energy and electricity markets. As a challenge for further research, the authors identify the consideration of unit commitment constraints with binary variables for the lower-level problem. This type of unit commitment constraints are common in scheduling of production and on-site energy systems, and thus, lead to challenging bilevel problems.

As reviewed, there are solution methods available for solving bilevel problems in general. However, an application of a solution algorithm to the bilevel optimization of batch production and energy systems scheduling is missing. The application would tackle the challenge identified by Wogrin et al. (2020).

2.3 Contribution of this thesis

The literature review in Section 2.1 shows that much progress has been made in the scheduling and synthesis of both energy systems and batch production systems. Additionally, the literature review shows the consideration of the energy demand has received increasing interest in production system scheduling. In Chapter 1, we pointed out that production scheduling is a major field in Process Systems Engineering and that the research field of Process Systems Engineering should be extended to other domains, such as Energy Systems Engineering (Pistikopoulos et al., 2021). Thus, combining energy system scheduling and synthesis with production system scheduling and synthesis merges importance concepts and methods from Energy and Production Systems Engineering. Nevertheless, for batch production and energy systems in industrial sites, optimization methods are missing. In particular, only a few papers deal with the integrated scheduling of batch production and energy systems. To the best of our knowledge, no research is available on Stackelberg games with either complete or incomplete information between energy and batch productions systems. Additionally, the integrated synthesis and control-reserve provision by batch production and energy system has also not been considered. However, methods that allow for efficiency increase in industrial sites with integrated energy and production systems are desired to decrease greenhouse gas emissions (c.f. Chapter 1).

To categorize the optimization methods proposed in this thesis, we differentiate between two possible relationships between the systems: Both systems are operated either by a single company, or both systems are operated by multiple companies (Figure 2.1). If a single company operates the systems, we assume that complete information is available and the systems optimize an overall objective function. Consequently, an integrated optimization should be applied.

If the systems are operated by multiple companies, we define the relationship as a Stackelberg game, and thus, assume a hierarchical structure in the decision-making process. The leader decides first, and the follower decides depending on the leader's decision. In an industrial site, first, the production system optimizes its operation to fulfill the product demand and reports to the energy system the corresponding energy demand. The energy system has to fulfill the energy demand. Thus, subsequently, the energy system optimizes its operation depending on the announced energy demand. Consequently, the production system is the leader, and the energy system is the follower.

The Stackelberg game can be further differentiated in the two cases where the production system has either incomplete or complete information on the energy system. If the production system has incomplete information on the energy system, coordination between the systems based on the available incomplete information can be performed. Thereby, the production system can optimize its objective while anticipating the behavior of the energy system with the incomplete information available. If the production system has complete information on the energy system, the production system can formulate a bilevel problem. In the bilevel problem, the production system can consider the reaction of the energy system exactly as the energy systems optimization problem is embedded in the production systems constraints.

For the integrated optimization and the Stackelberg-based optimization, this thesis provides optimization methods. The optimization methods are presented in two parts: In Part I, we present optimization methods for the integrated optimization, and in Part II, we provide optimization methods for Stackelberg-based optimization. A short outline of these optimization methods and their contribution are presented in the following as the contribution of this thesis.

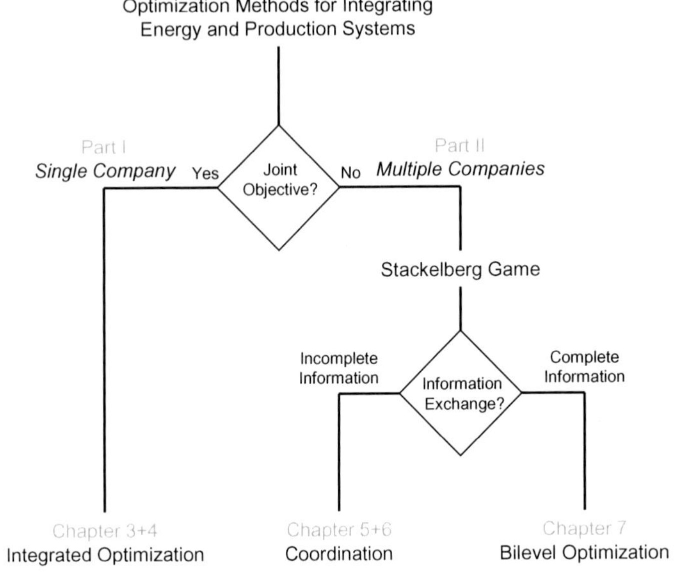

Figure 2.1: Contribution of this thesis: Optimization methods for the three possible relationships between energy and production systems.

Part I: Integrated optimization

Chapter 3: Integrated synthesis. For scheduling of energy and batch production systems, the integrated scheduling has been shown to be highly beneficial (Section 2.2.1). Thus, it seems desirable to expand the integrated scheduling to an integrated synthesis of both systems. Thereby, synergistic effects can already be considered during the synthesis phase, and the systems are built to operate more efficiently. For this purpose, we provide a method for the integrated synthesis of both systems that considers the synergistic effects of the integrated system. The proposed method integrates two optimization problems into an overall optimization problem.

Chapter 4: Integrated control-reserve provision. The increasing supply of electricity by renewable energy technologies increases the demand for control reserve for a stable electricity grid operation. Commonly, control reserve is provided by either energy systems or production systems, where mainly continuous operated production systems are considered. However, an integrated control-reserve provision by batch production and energy systems seems promising but has not been considered in literature so far. For this purpose, we provide a method for the integrated control-reserve provision. The method needs to consider the uncertain request of control reserve while considering that batch operations cannot be interrupted. For this purpose, we employ stochastic optimization to consider the upcoming uncertainties of control-reserve request while we set the production scheduling as non-anticipativity constraints to fix the production schedule.

Part II: Stackelberg-based optimization

Chapter 5: Coordination with incomplete information. In large industrial sites, energy and production systems are often operated by different companies. Thus, the two systems have different objectives and share incomplete information.
In the literature review in Section 2.2.2, we identified that Stackelberg games with incomplete information are applied to many applications to describe the interaction between an energy consumer and an energy supplier. We reviewed that the exchanged information between the two systems is commonly the energy demand and the energy price. However, for energy and batch production systems in industrial sites, no applications are available that solve the Stackelberg game with incomplete information.
One straightforward approach would be also to use the energy demand and the energy price as incomplete information. However, we show that this incomplete information does not well represent the complex cost structure of the energy system. Thus, the challenge is to identify incomplete information that leads to a good representation of the energy system's complex behavior. The incomplete information needs to satisfy the requirements of the energy system on confidentiality. For this purpose, we propose incomplete information that approximates the complex behavior of the energy system by only using energy cost for the current and a few additional energy demands. This incomplete information is then used in a method to solve the Stackelberg game between a single production and a single energy system.

Chapter 6: Coordination of multiple energy and production systems. In this chapter, we extend the method of Chapter 5 to multiple energy and production systems. Besides the coordination between the energy systems and the production systems, the challenge is to also coordinate among the energy systems and among the production systems. For this purpose, we propose novel coordination methods for the production systems and the energy systems. In particular, to assign the energy cost to the production systems, we present a method for allocating the demand-dependent energy cost. For the energy systems coordination, we propose an optimization problem for energy price minimization.

Chapter 7: Bilevel optimization. A drawback of the proposed methods to solve the Stackelberg game with incomplete information is that the optimal solution cannot be guaranteed. If the production system has complete information on the energy system, the optimal solution of the Stackelberg game can be guaranteed by solving a bilevel problem.
As reviewed in Section 2.2.3, solution algorithms are available to solve bilevel problems. However, no application schedules energy systems combined with batch production systems in bilevel problems. Furthermore, there is a need for solution algorithms with integer decisions in the lower-level problem to model the operational behavior of the energy system sufficiently. Thus, a solution algorithm is needed that solves the MILBP for energy and production system scheduling. For this purpose, we propose a solution algorithm to solve the bilevel problem between energy and production systems. The solution algorithm is an extension of the algorithm by Djelassi et al. (2019), which uses discretization points to represent the lower-level solution in the upper-level optimization problem.

Concluding, this thesis answers the question of how to optimize integrated energy and production systems in industrial sites, and proposes optimization methods in all the identified relationships between the systems.

Part I

Integrated optimization

Chapter 3

Optimal synthesis of integrated energy and batch production systems

In this chapter, we present a method for the integrated synthesis of energy and batch production systems. For this purpose, we extend the integrated scheduling of energy and batch production systems to the integrated synthesis. In integrated optimization, we assume a single company and thus optimize to a joint objective (Figure 2.1).

We combine a model for the synthesis of batch production systems with a model to synthesize energy systems. The chapter focuses on the combination of both problems since the method employs already existing models for the separated synthesis. We show the benefits of the integrated synthesis.

The chapter is organized as follows: In Section 3.1, the method and the corresponding model formulation are presented. In Section 3.2, the method is applied to two case studies. In both case studies, we show by parameter studies the benefits for industries with different shares of energy costs to overall costs. In Section 3.3, we draw conclusions.

Major parts of this chapter are reproduced by permission of Elsevier with modifications from the complete Elsevier source from:

Leenders, L., Bahl, B., Lampe, M., Hennen, M., and Bardow, A. (2019b). Optimal design of integrated batch production and utility systems. *Computers & Chemical Engineering*, 128, 496-511.

The author of this thesis contributed to the development and implementation of the method as well as the calculation and interpretation of the results, wrote the draft of the paper and is its principal author.

3.1 Integration of energy and production system synthesis

In this section, the method for the integrated synthesis of production systems and energy systems is presented. First, the conceptual overview is presented (Section 3.1.1). In the following, the detailed mathematical formulation of the optimization problem is given (Section 3.1.2).

3.1.1 Problem characterization and conceptual overview of the method

The synthesis of both energy and production systems is often based on superstructures (Frangopoulos et al., 2002; Barbosa-Póvoa, 2007). A superstructure represents all possible design options. An optimization procedure reduces the superstructure to the optimal system design.

Commonly, energy and production systems are synthesized sequentially (Chapter 2.1), resulting in separated superstructures (Figure 3.1, top). First, the production system is synthesized from a superstructure that includes production equipment, storage equipment and their connections. The solution of the corresponding optimization problem yields the optimal production system and its corresponding energy demand based on the simultaneously solved scheduling problem. The energy demand is used as an input parameter for the energy system synthesis. Second, the energy system is synthesized from a superstructure that includes potential energy conversion units and their connections. This sequential treatment leads to suboptimal solutions because the synthesis of the energy system is performed for a fixed energy demand and trade-offs between both systems are neglected.

The proposed method integrates the superstructures of energy and production systems into one joint superstructure (Figure 3.1, bottom). The integration of both superstructures combines the synthesis of energy and production systems into a single synthesis problem, which optimizes the design trade-offs between both systems to yield an overall optimal system. Since the energy system has to cover the energy demand of the production system, both systems are coupled by the energy demand.

The proposed method accounts for synthesis on three levels (Frangopoulos et al., 2002): On the topmost level, decisions are made on which equipment is built for energy and production system. On the second level, the sizing of selected equipment

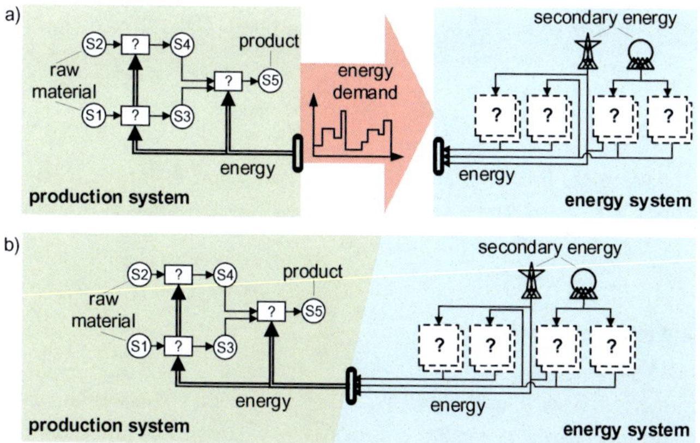

Figure 3.1: (a) Sequential synthesis: First, the production system (left) is optimized. The resulting energy demand of the production system is used as an input parameter for the optimization of the energy system (right). (b) Integrated synthesis: Energy and production system are coupled by the energy demand of the production system. The systems are optimized simultaneously. S: storage unit/state; ?: potential production equipment/energy conversion unit.

takes place. On the third level, the operational decisions are made for both the energy and production system.

In the following Section 3.1.2, the detailed mathematical model of the integrated superstructure is given.

3.1.2 Mathematical optimization model

In this section, a Mixed-Integer Linear Programming (MILP) model is presented for the integrated synthesis of energy and production systems. The production system is a batch production system. The production requires different energy forms which are supplied by the energy system. The objective is the minimization of the total annualized costs of the overall system $C^{\mathrm{TAC,TOTAL}}$. The overall costs $C^{\mathrm{TAC,TOTAL}}$ are the sum of the total annualized costs of the production system $C^{\text{TAC,PS}}$ and the total annualized costs of the energy system $C^{\text{TAC,ES}}$:

$$C^{\text{TAC,TOTAL}} = C^{\text{TAC,PS}} + C^{\text{TAC,ES}}. \tag{3.1}$$

In the following subsections, first, the models of the production system and energy system are described. Afterward, the connection between these systems is described.

3.1.2.1 Production system model

The modeled production system is a batch production system. The model of the batch production system is based on the *State-Task-Network-adapted* formulation from Pinto et al. (2008). The total annualized costs of the production system $C^{\text{TAC,PS}}$ are calculated by adding the annual operational expenditure $C^{\text{OPEX,PS}}$ to the capital expenditure $C^{\text{CAPEX,PS}}$ annualized by the present value annuity factor $\text{pvaf} = \frac{(1+\text{q})^{\text{n}}-1}{(1+\text{q})^{\text{n}}\cdot\text{q}}$ with the interest rate q and the number of periods n:

$$C^{\text{TAC,PS}} = \frac{C^{\text{CAPEX,PS}}}{\text{pvaf}} + C^{\text{OPEX,PS}}. \tag{3.2}$$

The capital expenditure $C^{\text{CAPEX,PS}}$ considers costs for each production equipment j, the storage units s, and the piping between production equipment j and storage unit s:

$$\begin{aligned} C^{\text{CAPEX,PS}} &= \sum_{j\in J} (V_j \cdot \text{C}_j^{\text{var}} + \gamma_j \cdot \text{C}_j^{\text{fix}}) + \sum_{s\in S} (V_s \cdot \text{C}_s^{\text{var}} + \gamma_s \cdot \text{C}_s^{\text{fix}}) \\ &+ \sum_{j\in J}\sum_{s\in S} (V_{s,j}^{\text{piping}} \cdot \text{C}_{s,j}^{\text{piping, var}} + \gamma_{s,j}^{\text{piping}} \cdot \text{C}_{s,j}^{\text{piping, fix}}). \end{aligned} \tag{3.3}$$

The capital expenditure takes into account economy of scale. Following the literature on batch production system synthesis (e.g., Pinto et al. (2008); Seid and Majozi (2013)), we model the economies of scale by a fixed cost factor C^{fix} for buying production equipment and a linear cost factor C^{var} multiplied with the equipment size. If required, the unit model could be refined to reflect economy of scale with more than one linear section. The capital costs of each unit increase linearly with a variable cost part C_j^{var}, C_s^{var}, $\text{C}_{s,j}^{\text{piping, var}}$ on the size of the unit V_j, V_s, $V_{s,j}^{\text{piping}}$. Each unit has a fixed cost part C_j^{fix}, C_s^{fix}, $\text{C}_{s,j}^{\text{piping, fix}}$ multiplied by the corresponding binary variable γ_j, γ_s, $\gamma_{s,j}^{\text{piping}}$ which identifies if a unit is bought.

The operational expenditure $C^{\text{OPEX,PS}}$ considers costs for the operation of production equipment j in time step t performing task i:

$$C^{\text{OPEX,PS}} = \frac{\text{t}^{\text{year}}}{\text{t}^{\text{cycle}}} \cdot \sum_{t \in T} \sum_{j \in J} \sum_{i \in I} \sum_{t'=t-\Delta\text{t}_{i,j}^{\text{process}}+1}^{t} (B_{t',i,j} \cdot \text{OC}_{i,j}^{\text{var}} + W_{t',i,j} \cdot \text{OC}_{i,j}^{\text{fix}}). \quad (3.4)$$

The operational expenditure $C^{\text{OPEX,PS}}$ consists of two parts: The first part depends linearly on the size of the processed batch $B_{t,i,j}$, multiplied by the variable operational costs $\text{OC}_{i,j}^{\text{var}}$ of the batch. The second part is the fixed cost part for processing the corresponding batch, which is modeled by the multiplication of the fixed costs $\text{OC}_{i,j}^{\text{fix}}$ for batch processing per time step and the binary variable $W_{t,i,j}$. The binary variable $W_{t,i,j}$ identifies if task i has started on production unit j in time step t. In Equation (3.4), the costs are summed from $t' = t - \Delta\text{t}_{i,j}^{\text{process}} + 1$ to t, where $\Delta\text{t}_{i,j}^{\text{process}}$ is the total time for executing task i on production equipment j, to consider fixed and variable operational costs for every time step the task is processed. To calculate the annual operational expenditure, the costs are multiplied by the annual production hours of the production system t^{year} divided by the hours of a production cycle t^{cycle}.

For every considered energy form e (e.g., heat, cold, electricity), the energy demand of the batch production system is calculated in each time step t:

$$E_{t,e}^{\text{demand}} = \sum_{j \in J} \sum_{i \in I} \sum_{t'=t-\Delta\text{t}_{i,j}^{\text{process}}+1}^{t} (B_{t',i,j} \cdot V_{t-t'+1,i,j,e}^{\text{var}} + W_{t',i,j} \cdot V_{t-t'+1,i,j,e}^{\text{fix}}) \quad \forall t \in T, e \in E. \quad (3.5)$$

We consider a fixed $V_{t,i,j,e}^{\text{fix}}$ and a variable $V_{t,i,j,e}^{\text{var}}$ energy demand of the tasks.

3.1.2.2 Energy system model

The model of the energy system is based on Voll et al. (2013) and considers multiple types of energy conversion units u and energy forms e.

As for the production system, the total annualized costs of the energy system $C^{\text{TAC,ES}}$ are also calculated as the sum of the annual operational expenditure $C^{\text{OPEX,ES}}$ and the capital expenditure $C^{\text{CAPEX,ES}}$, which is annualized by the present value annuity factor pvaf:

$$C^{\text{TAC,ES}} = \frac{C^{\text{CAPEX,ES}}}{\text{pvaf}} + C^{\text{OPEX,ES}}. \quad (3.6)$$

The capital expenditure of all energy conversion units $u \in U$ are summed to calculate the capital expenditure of the energy system $C^{\text{CAPEX,ES}}$:

$$C^{\text{CAPEX,ES}} = \sum_{u \in U} \sum_{h \in H_u} (\kappa_{u,h} \cdot \mathrm{C}^{\text{fix}}_{u,h} + V^{\text{N}}_{u,h} \cdot \mathrm{C}^{\text{var}}_{u,h}). \tag{3.7}$$

Economy of scale is represented by piecewise linearization of the investment cost curves for the energy conversion units. The linear sections h of the investment cost curve are specified by a fixed cost part $\mathrm{C}^{\text{fix}}_{u,h}$ multiplied by the binary variable $\kappa_{u,h}$, and a variable cost part $\mathrm{C}^{\text{var}}_{u,h}$ scaled by the nominal output of the energy conversion unit $V^{\text{N}}_{u,h}$. The binary variable $\kappa_{u,h}$ declares if an energy conversion unit is bought in section h ($\kappa_{u,h} = 1$) or not ($\kappa_{u,h} = 0$).

The annual operational expenditure of the energy system $C^{\text{OPEX,ES}}$ is calculated as the sum over all time steps t:

$$C^{\text{OPEX,ES}} = \frac{\mathrm{t}^{\text{year}}}{\mathrm{t}^{\text{cycle}}} \cdot \sum_{t \in T} \sum_{e \in E} \left[\Delta\mathrm{t} \cdot \left(P^{\text{buy}}_{t,e} \cdot \mathrm{p}^{\text{buy}}_{e} - P^{\text{sell}}_{t,e} \cdot \mathrm{p}^{\text{sell}}_{e} \right) \right] + \sum_{u \in U} CM_u. \tag{3.8}$$

The annual operational expenditure $C^{\text{OPEX,ES}}$ takes into account costs for purchased secondary energy $P^{\text{buy}}_{t,e}$ (e.g., electricity and gas) at a price $\mathrm{p}^{\text{buy}}_{e}$, revenues from energy sold to the grid $P^{\text{sell}}_{t,e}$ (e.g., electricity) at a price $\mathrm{p}^{\text{sell}}_{e}$ and costs for maintenance CM_u. The costs for maintenance CM_u are specified as fraction of the capital expenditure for each energy conversion unit. The detailed modeling is given in Appendix A.1.2.

The overall supply from the energy system $E^{\text{supply}}_{t,e}$ is calculated for each energy form e by:

$$E^{\text{supply}}_{t,e} = \sum_{u \in U} V_{t,u,e} - \sum_{u \in U} U_{t,u,e} - P^{\text{sell}}_{t,e} + P^{\text{buy}}_{t,e} \quad \forall t \in T, e \in E. \tag{3.9}$$

The sum of all energy supplied from the energy conversion units u is decreased by the energy flow to run other energy conversion units $U_{t,u,e}$ and the energy that is sold to the grid $P^{\text{sell}}_{t,e}$. Furthermore, the supplied energy is increased by the energy that is purchased from the grid $P^{\text{buy}}_{t,e}$.

The model of the energy conversion units ($V_{t,u,e}$ and $U_{t,u,e}$) is based on linearized part-load efficiencies and constraints for the minimal part load. The modeling of part-load is important for the operation and design optimization of energy systems. For example, compression chillers typically have their highest efficiency when running in part-load operation. Consequently, when neglecting part-load performance, the design would possibly be suboptimal in practice. The detailed model formulation is given in Appendix A.1.2.

3.1.2.3 Integration of energy system and production system

Both subsystems, the production system and the energy system, are connected by the energy demand of the production system (Figure 3.1, bottom). The energy demand of the production system $E_{t,e}^{\text{demand}}$ (Equation (3.5)) has to be satisfied by the energy supplied by the energy system $E_{t,e}^{\text{supply}}$ (Equation (3.9)) in every time step t and for all energy forms e. Thus, both systems are coupled by the energy balance:

$$E_{t,e}^{\text{demand}} = E_{t,e}^{\text{supply}} \quad \forall t \in T, e \in E. \tag{3.10}$$

The integrated optimization of both systems enables potential synergies between the systems. Thus, the proposed method for integrated synthesis can lead to different equipment and unit operations compared to the common sequential synthesis approach.

3.2 Case studies

The proposed method for integrated synthesis of production system and energy system is applied to two case studies that have been adapted from literature. The proposed method is compared to the common sequential synthesis approach. Other authors have also used a sequential procedure to benchmark the integrated scheduling of energy and production systems (Agha et al., 2010; Zhao et al., 2014; Zulkafli and Kopanos, 2016). In the sequential synthesis, first, the production system is synthesized. Afterward, the energy system is synthesized for the resulting energy demand. The models used in the sequential synthesis are the same as in the integrated synthesis, but without the coupling constraints (Equation (3.10)). The superstructures of the overall systems are presented in two separated illustrations each to improve readability (Figure 3.2, 3.3, 3.9, and 3.10). All production systems in this thesis are based on Kondili et al. (1993) and Kallrath (2002) and are modeled based on the STN formulation. All energy systems are modeled based on Voll et al. (2013) and Baumgärtner et al. (2019a). The model from Baumgärtner et al. (2019a) extends the model by Voll et al. (2013) with renewable energy conversion technologies.

The energy systems considered in this thesis are decentralized multi-energy systems. We classify the production systems and the energy systems used in this thesis by the

Table 3.1: Classification of energy and production system in the case studies with the triplet $\alpha|\beta|\gamma$. α: machine environment, β: characteristics, γ: objective function.

triplet	energy system	production system
α	flexible job shop	flexible job shop
β	continuous operation, jobs released and finished in each time step, mixing and splitting of material, no transportation time	batch operation, jobs released in first time step, fixed processing times, mixing and splitting of batches, no transportation time
γ	single objective (cost minimization or profit maximization)	single objective (cost minimization or profit maximization)

triplet $\alpha|\beta|\gamma$ (Table 3.1 and Section 2.1.1). The energy and the production system can be described as a flexible job shop. In a flexible job shop, each job has its predefined production route of steps. If necessary, a job may have to visit a machine multiple times. Flexible means that multiple machines can perform the same step. Thus, for each step, the scheduler also has to assign a machine for the job. The detailed topology of the production systems is described in the corresponding case study.

For modeling the production systems, we use the time representation that best fits to the energy system model, i.e., discrete-time representation and global time intervals. In general, the behavior of the production units might follow nonlinear performance curves. Commonly, in scheduling of production systems, the performance of the production units is modeled to be piecewise-affine. We also model piecewise-affine performance curves. Thus, we consider MILPs like most scheduling problems (Castro et al., 2018).

For modeling the energy systems, we use a discrete-time representation and global time intervals. In general, the performance of energy conversion units is nonlinear (Goderbauer et al., 2016). The nonlinear behavior is commonly approximated by piecewise-affine functions (Kotzur et al., 2020). We model the energy conversion units of the energy systems by piecewise-affine functions.

The optimization problems are formulated in GAMS 24.7.3 (GAMS Development, 2020) and solved with CPLEX 12.6.4.0 (IBM Corporation, 2020). The problems are solved on an Intel(R) Xeon (R) CPU E5-1660-v3 @ 3.00 GHz applying 6 threads and 32 GB RAM. All optimization problems are solved with a gap of 0.1 % and a time limit of 24 h (86,400 s). For the parameter studies of both case studies (Section

3.2.1.2 and 3.2.2.2), the time limit is reduced to 6 h (21,600 s). The optimization problem of case study I has 1067 discrete variables, 6315 continuous variables, and 12,050 constraints with 29,982 nonzero elements. The optimization problem of case study II has 11,593 discrete variables, 30,706 continuous variables, and 91,294 constraints with 203,227 nonzero elements.

In Section 3.2.1.1 and 3.2.2.1, results are shown for an energy cost share of around 40 % to the overall cost. In Section 3.2.1.2 and 3.2.2.2, the energy demands of the production systems are varied to analyze the benefits of the proposed method for industries with a high and low share of energy costs.

3.2.1 Case study I

The production system used in case study I is based on the work of Kondili et al. (1993) (Figure 3.2). This example has been extensively studied in literature for batch production system scheduling. Here, the example is extended to a synthesis problem and energy demands of the energy forms heat, cold and electricity are added (Table B.3).

The production system runs $t^{year} = 300$ days a year in a $t^{cycle} = 12$ hour cycle. The total annualized costs of the overall system $C^{TAC,TOTAL}$ are calculated for a time horizon of $n = 5$ years with an interest rate of $q = 4\%$. While production system lifetime is much longer, shorter time horizons are often assumed in practice for investment decisions to ensure fast payback time. The two objective functions of the sequential synthesis are the total annualized costs of the production system $C^{TAC,PS}$ (Equation (3.2)) and the energy system $C^{TAC,ES}$ (Equation (3.6)). For each 12 hour cycle, a final product demand of 105 t of S7 and 202.5 t of S10 needs to be satisfied. The production system produces no byproducts, nor are alternative production routes considered. Thus, because the demand for final products is fixed, the amount of raw materials and the revenues from selling final products are also fixed. The cost and revenues for these items are thus fixed and are not taken into account in our objective function (Equation (3.2)) since they do not influence the results. However, if desired, these costs could be added transforming the proposed objective function $C^{TAC,TOTAL}$ to total annualized profit $P^{TAP,TOTAL}$ (Equation (3.12)).

The superstructure and model of the energy system are based on Voll et al. (2013). The superstructure of the energy system considers 2 boilers, 6 combined-heat-and-power engines, 3 compression chillers and 3 absorption chillers (Figure 3.3 and Table 3.2). Electricity can be sold to the grid for $p_{el}^{sell} = 0.1$ €/kWh and can be purchased

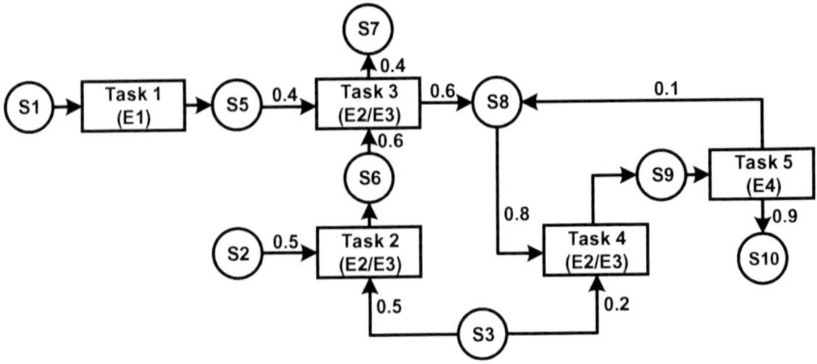

Figure 3.2: Superstructure of the production system in case study I based on Kondili et al. (1993). Here, we added energy demands (Table B.3). The values next to the arrows describe the production and consumption fraction of each task. These fractions declare how much of the corresponding state is needed or produced by a batch. S: storage unit/state; E: production equipment.

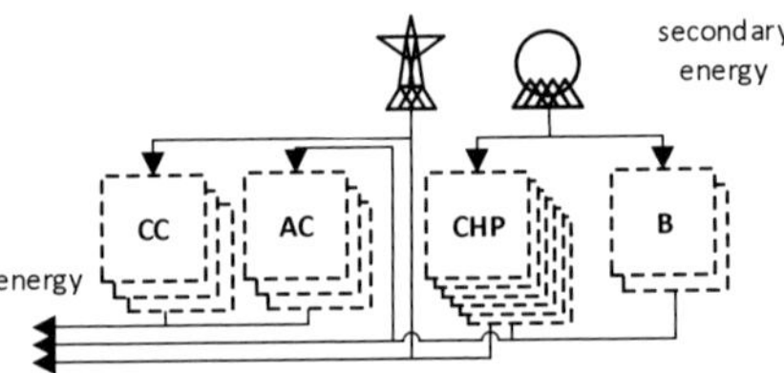

Figure 3.3: Superstructure of the energy system in case study I. The available capacities are given in Table 3.2. B: boiler; CHP: combined-heat-and-power engine; CC: compression chiller; AC: absorption chiller.

for $p_{el}^{buy} = 0.16$ €/kWh. Gas can be bought for $p_{gas}^{buy} = 0.06$ €/kWh. Additional parameters are provided in Appendix B.1.

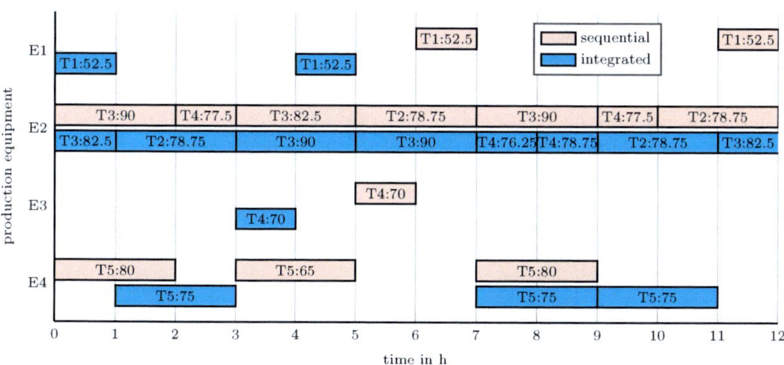

Figure 3.6: Gantt chart for the sequential synthesis and the integrated synthesis in case study I. Declaration: task name: batch size/tons; E: production equipment.

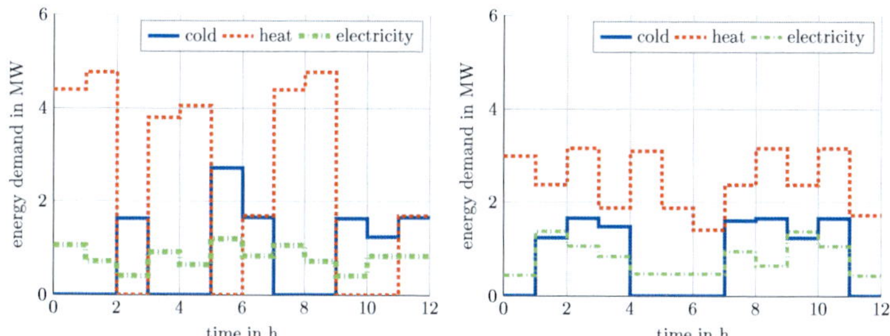

Figure 3.7: Energy demand curve of the optimal solution for the sequential synthesis (left) and for the integrated synthesis (right) in case study I.

steps the batches are processed. For example, tasks T3 and T5 are performed at the same time steps in the sequential synthesis (Figure 3.6), which results in high peak demands for heat (Figure 3.7). In the integrated synthesis, tasks T3 and T5 are never performed at the same time steps (Figure 3.6), and thus, the peaks in the heat demand are significantly reduced (Figure 3.7). These differences lead to an optimized energy demand curve with lower peaks and flatter valleys in the integrated synthesis (Figure 3.7), which finally leads to a smaller energy system. In detail, in the sequential synthesis, the highest heat demand is 4.77 MW. In the integrated synthesis, the highest heat demand is only 3.16 MW. Thus, the nominal heat power of the energy system is reduced significantly (Figure 3.5). The highest cooling demand is the same in both designs (1.65 MW). However, in the sequential synthesis, a cooling capacity of the energy system of 2.72 MW is installed with one absorption and one compression chiller. The integrated synthesis can avoid this overcapacity, where the cooling demand is changed such that the cooling capacity of the optimal energy system equals the highest cooling demand. The highest electricity demand is 16 % higher in the integrated synthesis (1.38 MW to 1.19 MW). The lower peak demands in the heat and cooling demand lead to significant cost savings in the investment costs in the integrated synthesis (40 %, Figure 3.4).

In the sequential synthesis, the production system optimization reaches the gap of 0.1 % after 385 s, the subsequent energy system optimization is solved after 14 s. The integrated synthesis reaches the time limit of 24 h with a remaining gap of 1.72 %. However, the integrated synthesis finds a better solution than the sequential synthesis within 2 s. After 15 s, a solution is found with an estimated gap of 6.11 %. With knowledge of the final solution, we know that this solution actually has a gap of only 1.97 %. The final solution of the integrated synthesis is found after 10.75 h with a remaining gap of 1.83 %. Thus, the computational effort is significantly higher for the integrated synthesis, but significantly better solutions are found fast and the computational effort mainly goes into the optimality proof. The computation time seems to be limiting in practice for studies like a detailed sensitivity analysis. However, the integrated synthesis has been shown to find better results than the sequential synthesis within a few seconds. Thus, a sensitivity analysis on good solution candidates seems practicable. To speed up the computation, we tested different solver settings but with little effect. If the problem is to be solved repeatedly, the development of problem-specific solution strategies could be promising.

In this case study, the energy costs have a share of 40 % of the overall cost. In the following Section 3.2.1.2, the energy cost share is varied by a parameter study to identify the benefits of the integrated synthesis for different industry sectors.

3.2.1.2 Results of the parameter study: variation of the energy cost share

The cost savings of the integrated synthesis depend on the energy cost share. To quantify the impact of the energy cost share, case study I (Figure 3.2) is recomputed with varied energy demands of the production tasks. For this purpose, the energy demand is multiplied by a factor between 0.125 and 3. All other parameters stay the same. Thus, the energy cost share varies. However, the exact variation of the energy cost share cannot be predicted and will depend on the optimization results. This

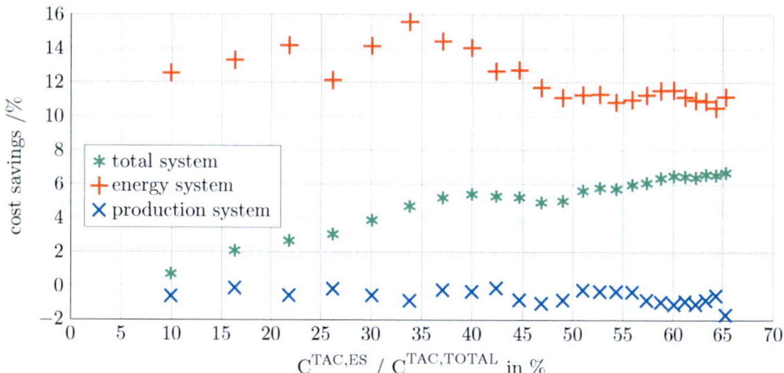

Figure 3.8: Cost savings by the integrated synthesis compared to the sequential synthesis for case study I for different cost shares of the energy systems $C^{\mathrm{TAC,ES}}/C^{\mathrm{TAC,TOTAL}}$ in the total annualized costs of the total system. The individual cost savings of the energy and production system are related to their own cost each.

variation of the energy demand can be used to estimate the benefits of the integrated synthesis for different industries with different energy cost shares to the overall cost. In the parameter study, the sequential synthesis does never exceeds the time limit. In contrast, the integrated synthesis always reaches the time limit and the remaining gap varies from 0.4 % to 4.1 %.

As expected, the results show that the cost benefits of the integrated synthesis rise with increasing energy cost shares (Figure 3.8). The overall cost savings reach values up to 6.7 %. The cost for the production system increases up to 2 %. Thus, only small additional costs are caused in the production system, which are always more than compensated by cost savings of the energy system. The energy cost savings are always higher than 10 % and reach values up to almost 16 %. The savings in the energy system indicate the great benefit of the integrated synthesis. Even if energy costs represent only a small share of the total cost, i.e., starting from 15 % upwards, the integrated synthesis of energy and production systems can always save over 2 % of the total cost.

3.2.2 Case study II

The production system used in case study II is based on the work of Kallrath (2002) (Figure 3.9). This example has also been extensively studied in literature for batch production system scheduling. Here, we extend the example to a synthesis problem. The cost as well as minimum and maximum capacities of the production system are adapted from Pinto et al. (2008). To consider the energy system, we added energy demands for heat, cold, and electricity (Table B.6).

The production system runs $\mathrm{t}^{\text{year}} = 300\,\text{days}$ per year. In this case study, we consider the revenues and cost for selling products and buying raw materials, respectively. Thus, the set of variables now also includes the amount and type of products $product_{t,s}$ and the bought raw material $raw_{t,s}$. Consequently, the annual operational expenditure $C^{\text{OPEX,PS}}$ are here given by:

$$C^{\text{OPEX,PS}} = \frac{\mathrm{t}^{\text{year}}}{\mathrm{t}^{\text{cycle}}} \cdot \Bigg(\sum_{t \in T} \sum_{j \in J} \sum_{i \in I} \sum_{t'=t-\Delta \mathrm{t}_{i,j}^{\text{process}}+1}^{t} (B_{t',i,j} \cdot \mathrm{OC}_{i,j}^{\text{var}} + W_{t',i,j} \cdot \mathrm{OC}_{i,j}^{\text{fix}}) + \sum_{t \in T} \sum_{s \in S} (\mathrm{p}_s \cdot raw_{t,s} - \mathrm{p}_s \cdot product_{t,s}) \Bigg). \quad (3.11)$$

Compared to the operational expenditure in Equation (3.4), we also consider cost and revenues for purchasing and selling. For the raw material S1 with the amount $raw_{t,s}$, the price p_s is 2 €/t. For the products $product_{t,s}$ S15-S19, the price p_s is 150 €/t. Products can only be sold at the end of the production time $\mathrm{t}^{\text{cycle}} = 36\,\text{h}$.

The objective function in case study II is the total annualized profit $P^{\text{TAP,TOTAL}}$ which corresponds to the negative of the total annualized cost $C^{\text{TAC,TOTAL}}$

$$P^{\text{TAP,TOTAL}} = -C^{\text{TAC,PS}} - C^{\text{TAC,ES}}. \quad (3.12)$$

The total annualized profit of the overall system $P^{\text{TAP,TOTAL}}$ is calculated for a time horizon of $\mathrm{n} = 5\,\text{years}$ with an interest rate of $\mathrm{q} = 4\,\%$. The two objective functions of the sequential synthesis are the total annualized costs of the production system $C^{\text{TAC,PS}}$ (Equation (3.2)) where we use now the $C^{\text{OPEX,PS}}$ from Equation (3.11) and the total annualized costs of the energy system $C^{\text{TAC,ES}}$ from Equation (3.6). The superstructure of the energy system considers 3 boilers, 3 combined-heat-and-power engines, 3 compression chillers, and 3 absorption chillers (Figure 3.10 and Table 3.2). Electricity can be sold to the grid for $\mathrm{p}_{el}^{\text{sell}} = 0.1$ €/kWh and can be purchased for

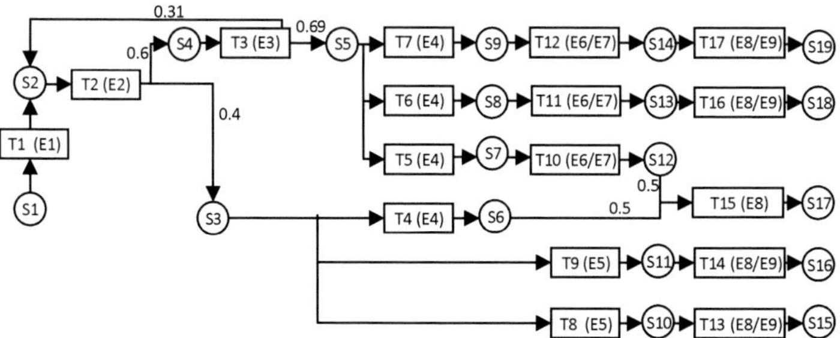

Figure 3.9: Superstructure of the production system in case study II based on Kallrath (2002). Here, we added energy demands (Table B.6). The values next to the arrows describe the production and consumption fraction of each task. These fractions declare how much of the corresponding state is needed or produced by a batch. S: storage unit/state; E: production equipment; T:task.

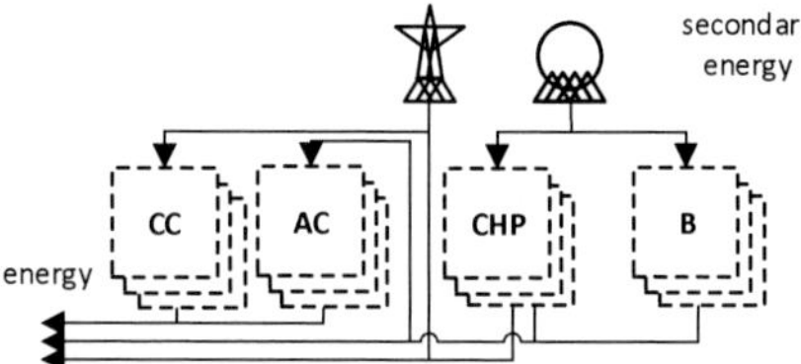

Figure 3.10: Superstructure of the energy system in case study II. The available capacities are given in Table 3.2. B: boiler; CHP: combined-heat-and-power engine; CC: compression chiller; AC: absorption chiller.

$\mathrm{p}_{el}^{\mathrm{buy}} = 0.16$ €/kWh. Gas can be bought for $\mathrm{p}_{gas}^{\mathrm{buy}} = 0.06$ €/kWh. Additional parameters are provided in Appendix B.2.

3.2.2.1 Results for the base energy demand

In case study II, the integrated synthesis increases the total profit by 5.5 % compared to the sequential synthesis (Figure 3.11). The profit increases due to the cost decrease for the energy supply $C^{\mathrm{TAC,ES}}$ by 11.9 %. In contrast, the total annualized costs for the production system without the cost for sold and purchased material $C^{\mathrm{TAC,PS}}$ (Equation (3.2)) increase by 2.6 %. While the sequential and the integrated synthesis both

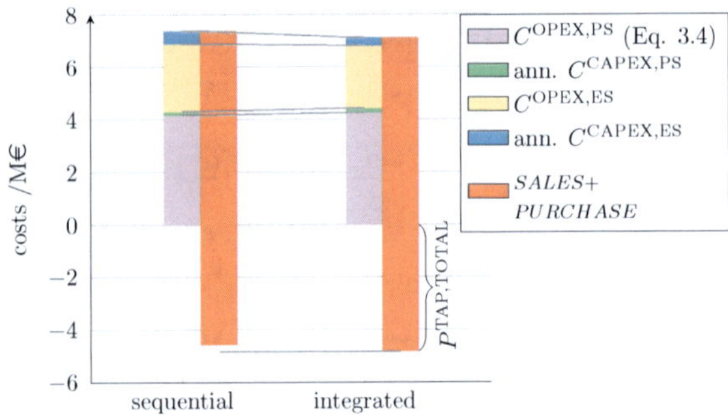

Figure 3.11: Costs for the sequential synthesis and the integrated synthesis in case study II. *OPEX*: operational expenditure; *CAPEX*: capital expenditure; *P*: profit; ann: annualized.

purchase the same amount of raw material (S1: 525.6 t), their products differ: Both designs produce 331.2 t of S17, but the sequential design additionally produces 74.4 t of S16 while the integrated design produces 74.4 t of S15. However, all products have the same specific prices and the same quantity of raw materials is bought. Thus, both approaches have the same revenues from selling products and buying raw materials. Still, total annualized profit is increased significantly by performing different tasks and producing different products (Figure 3.11).

The resulting designs differ (Figure 3.12): In the integrated synthesis, all production equipment is greater or at least equal to the sequential synthesis: The increase is 50 % for equipment E3 and E6 and 52 % for equipment E4 and E8. In contrast, in the sequential synthesis, the energy system is larger with one additional combined-heat-and-power engine. Furthermore, several units are larger (Figure 3.12): boiler B1 by 27 %, boiler B2 by 182 %, compression chiller CC1 by 87 %, compression chiller CC2 by 93 %, and absorption chiller AC by 12 %.

The Gantt chart differs significantly between the integrated synthesis and the sequential synthesis (Figure 3.13). Different tasks produce different products. In the integrated synthesis, equipment E3, E4, E6, and E8 are larger to perform their tasks with a larger maximal batch size. Consequently, the resulting energy demand profiles differ significantly (Figure 3.14): The peak demand of the electricity demand increases slightly by 4.1 % in the integrated synthesis. However, the peaking heating and cooling demand are reduced significantly in the integrated synthesis by 37 % and 46 %,

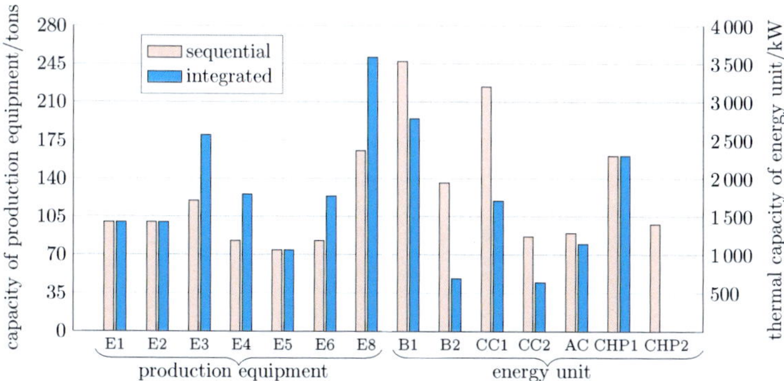

Figure 3.12: Optimal capacity of production equipment and energy conversion units for the integrated and the sequential synthesis in case study II. E: production equipment; B: boiler; CC: compression chiller; AC: absorption chiller; CHP: combined-heat-and-power engine.

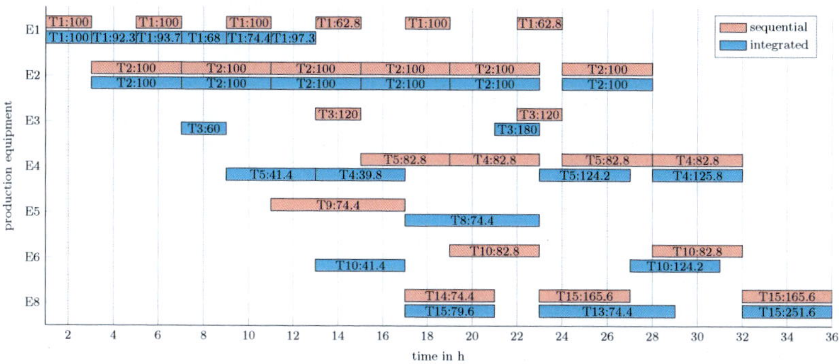

Figure 3.13: Gantt chart for the sequential synthesis and the integrated synthesis in case study II. Declaration: task name; batch size/tons. E: production equipment.

respectively. Thereby, a smaller energy system is sufficient in the integrated synthesis as stated before (Figure 3.12).

In the sequential synthesis, the production system optimization reaches the gap of 0.1 % after 662.7 s, the subsequent optimization of the energy system is solved in 569.6 s. The integrated synthesis reaches the time limit of 24 h with a remaining gap

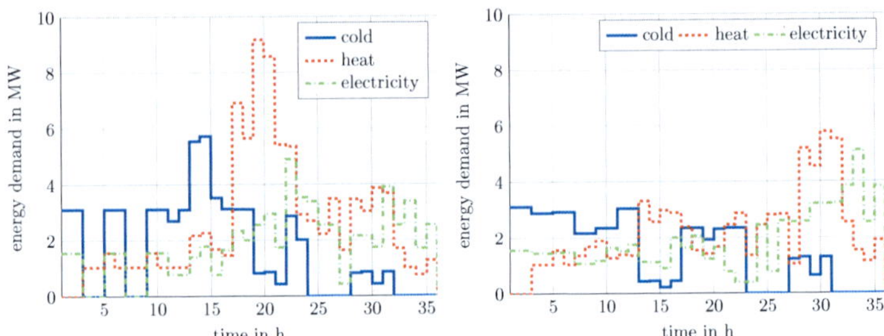

Figure 3.14: Energy demand curve of the optimal solution for the sequential synthesis (left) and for the integrated synthesis (right) in case study II.

of 4.0 %. However, the integrated synthesis finds a better solution than the integrated synthesis within 86.4 s. After 600 s, a solution is found with an estimated gap of 8.4 %. With knowledge of the final solution, we know that this solution has a gap of only 4.7 %. Thus, the computational effort is higher for the integrated synthesis. Still, significantly better solutions are found fast, even for such large optimization problems, and the computational effort mainly goes into the optimality proof.

In case study II, the costs of the energy system have a share of 38 % of the total annualized cost $C^{\mathrm{TAC,TOTAL}}$. In the following Section 3.2.2.2, the energy cost share is varied by a parameter study to identify the benefits of the integrated synthesis for different industry sectors.

3.2.2.2 Results of the parameter study: variation of the energy cost share

The increase in total profit of the integrated synthesis depends on the energy cost share, as shown in case study I. To quantify the impact of the energy cost share on the increase in total profit also for case study II (Figure 3.9 and 3.10), we multiplied the energy demands of the production tasks by a factor between 0.25 and 2.5. All other parameters stay the same.

For energy demands multiplied with a factor of 1.5 or higher, the sequential synthesis reaches the given time limit before the gap 0.1 % is met. The remaining gap is up to 1.8 %. The integrated synthesis always reaches the time limit. The remaining gap varies from 3.3 % to 7.7 % and is even higher for the three cases with the highest multiplication factors.

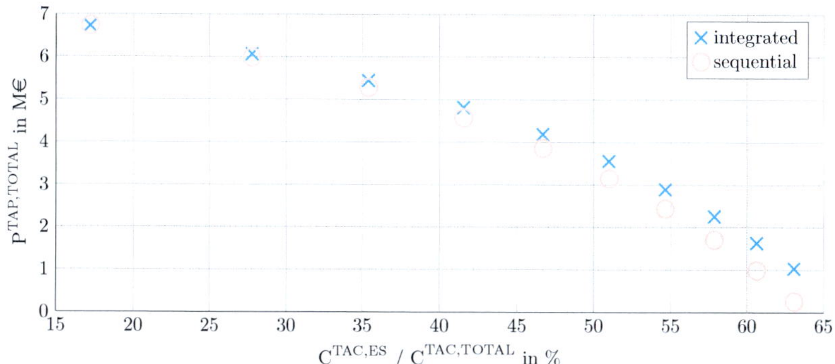

Figure 3.15: Total annualized profit by the integrated synthesis and the sequential synthesis for case study II for different cost shares of the energy systems $C^{\mathrm{TAC,ES}}/C^{\mathrm{TAC,TOTAL}}$ in the total annualized costs of the overall system.

As in case study I, the benefit from the integrated synthesis rises with an increasing energy cost share (Figure 3.15). The total profit increases up to of 299 % for the highest energy cost share. However, this increase is caused by the low total profit in the sequential synthesis. For lower energy cost shares, the increase in total profit reaches values up to 66 %. The sequential synthesis always produces the same products, while the integrated synthesis adapts its products. For example, 34.4 t of S15 and 331.2 t of S17 are produced for the highest energy demand, while 74.4 t of S16 and 331.2 t of S17 are produced for the lowest energy demand. For the lowest energy cost share, the integrated synthesis decreases total profit by 0.1 %, due to the short computation time in the parameter study and the remaining gap of 4.4 %.

3.3 Conclusions

Energy and production systems are commonly synthesized sequentially. The sequential synthesis leads to suboptimal designs regarding the overall objective of both systems. Thus, in this chapter, a method is proposed for the integrated synthesis of energy and production systems. Both systems are coupled by the energy demand of the production system, which has to be supplied by the energy system. The integration of both optimization problems leads to an overall superstructure, which allows to optimize the design trade-off between the production system and energy system to an overall optimal system.

The integrated synthesis is compared to a sequential synthesis for two case studies from the scheduling literature. In case study I, the product demand is predefined for the time horizon, and the integrated synthesis saves 14 % in the total annualized costs of the energy system compared to the sequential synthesis. The total annualized costs for the production system rise by 0.3 %, but the overall total annualized costs of both systems are still reduced by 5.4 %. The results show that the optimal structure and schedule of the production system differ in the sequential and the integrated synthesis. The optimal energy system needs fewer energy conversion units to supply the required energy to the production system. Furthermore, through the integrated synthesis, peak demands of heat and cold are reduced. We show the benefits of the integrated synthesis for different energy cost shares in a parameter study.

In case study II, the amount and type of products are added as decision variables, and the total profit is optimized. The integrated synthesis increases the total profit by 5.5 % and produces different products compared to the sequential synthesis. As for case study I, a parameter study is provided, which shows the increased benefit of the integrated synthesis for an increasing share of energy cost to the total annualized cost.

The case studies show that the integrated synthesis of energy and production systems leads to significant cost savings of the overall system. Cost savings in the energy system are high, with only small additional costs for the production system. The energy system needs fewer energy conversion units, and thus, resources for the construction are saved.

In general, integrated optimization is not only desirable in the synthesis but also during the operation of the systems. For operational optimization of energy and production systems, more complex models can be considered compared to the synthesis because the design decisions are already fixed. More complex models can consider additional participation in electricity and balancing-power markets. In particular, the provision of control reserve can be considered by an integrated optimization to decrease operational costs. Thus, in the following chapter, we present a method to optimize the operation for an integrated control-reserve provision by energy and production system.

Chapter 4

Integrated scheduling of energy and batch production systems for provision of control reserve

In this chapter, we propose a method for the integrated scheduling of energy and batch production systems for control-reserve provision. As in Chapter 3, we assume a single company, and thus, optimize to a joint objective (Figure 2.1). However, compared to the method in Chapter 3, we fix the design decisions and extend the model by participation in the control-reserve market. The challenge for control-reserve provision by batch production systems is that batches can often neither be interrupted nor run in part-load. Thus, the proposed method identifies a production schedule that remains fixed during a request of control reserve. Still, the fixed production schedule leads to an energy demand that increase profits from control-reserve provision.

The chapter is organized as follows: In Section 4.1, we extend the model from Chapter 3. In Section 4.2, the method is applied to a case study where the integrated system participates in control-reserve market. Therein, we show the changes in the energy systems operation if control reserve is requested. In Section 4.3, we conclude.

Major parts of this chapter are reproduced by permission of the ECOS 2020 Local organizing committee with modifications from the complete ECOS source from:

Leenders, L., Starosta, A., Baumgärtner, N., and Bardow, A. (2020). Integrated scheduling of batch production and utility systems for provision of control reserve. In *Proceedings of ECOS 2020 - 33rd International Conference on Efficiency, Cost, Optimization, Simulation and Environmental Impact of Energy Systems*, pages 712-723. Osaka, Japan.

The author of this thesis contributed to the development and implementation of the method as well as the calculation and interpretation of the results, wrote the draft of the paper and is its principal author.

4.1 Integrated scheduling for provision of control reserve

In this section, we present a method for scheduling energy and batch production systems to participate in the control-reserve market. The method is intended for the frequent scheduling of production systems, e.g., daily production scheduling.

In this method, we overcome the sequential scheduling by simultaneously scheduling energy and production systems while optimizing the participation of the energy system in the control-reserve market (Figure 4.1). Our method assumes that the requested control-reserve energy is provided by changing the operation of the energy system since the production schedule is fixed. The integrated scheduling identifies the optimal production schedule for the provision of control reserve by the energy system.

Our method takes the probability of control-reserve requests into account in a stochastic optimization problem with three scenarios: 1. Positive control reserve is requested (POS), 2. Negative control reserve is requested (NEG), 3. No control reserve is requested (NO).

In Section 4.1.1, we state the problem of providing control-reserve in control-reserve markets and explain the stochastic process of the considered problem. In Section 4.1.2, we model the integrated scheduling with control-reserve provision. Here, we focus on the extensions compared to the modeling without control-reserve provision (Chapter 3, note that we omit design decisions in this chapter).

4.1.1 Stochastic problem of control-reserve provision

The method uses a stochastic programming model for participation in the control-reserve market. Here, we consider the control-reserve market as a pay-as-bid market, e.g., the German and French tertiary control-reserve markets (Frontier Economics, 2016).

The method considers the provision of both positive and negative control reserve. Positive control reserve is requested if an insufficient amount of electricity is supplied to the grid. Consequently, the electricity supply needs to be increased or the electricity demand decreased. If negative control reserve is requested, electricity supply needs to be decreased, or electricity demand needs to be increased.

Our method considers a control-reserve market where offers contain two prices: the capacity price for providing control-reserve capacity and the energy price for the delivery of control-reserve energy. The offers of all market participants are ordered by the capacity price in a merit order. All offers are accepted in the merit order until

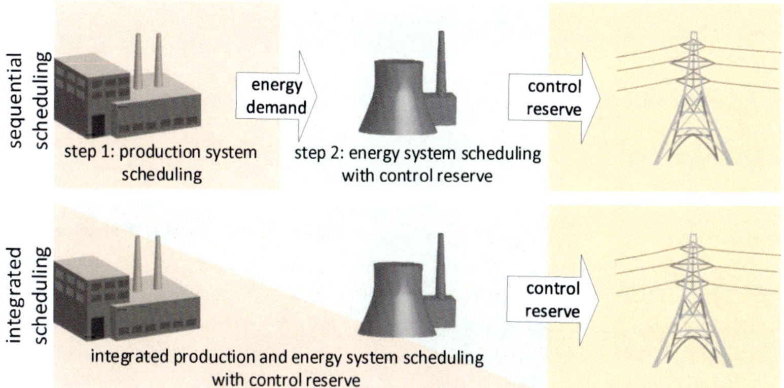

Figure 4.1: Sequential and integrated scheduling of energy and production system. In sequential scheduling, first, the production system is scheduled for the minimal production cost. Subsequently, the energy system is scheduled to provide the energy demand of the production system as well as control reserve to the grid. The integrated scheduling considers the energy and production system simultaneously.

the demand of control-reserve capacity is met. The offered capacity price is paid for every accepted offer. Subsequently, the accepted offers are ordered by their energy prices in a second merit order. If control reserve is needed, control reserve is activated in the order of the second merit order. These market mechanics correspond, e.g., to the tertiary control-reserve market in Germany (50Hertz Transmission GmbH et al., 2019).

Our method optimizes the participation of an integrated energy and batch production system in a control-reserve market by optimizing the offered control-reserve capacity and the energy price for the actually delivered control reserve. The requested control reserve is delivered by adapting the operation of the energy conversion units while the electricity demand or supply to the grid meets the control-reserve request. To identify the optimal offer, we take into account the probability of control-reserve requests. The probability of a control-reserve request depends on the offered energy price. Higher energy price offers lead to lower probabilities of control-reserve requests and vice versa.

We model the uncertainty of request by a two-stage stochastic programming model (Birge and Louveaux, 2011). Modeling the participation in control-reserve markets by stochastic programming is a common approach (Kumbartzky et al., 2017; Schäfer et al., 2019; Bohlayer et al., 2020). Compared to the literature reviewed in Section 2.1.2 and Section 2.1.3, we model the provision of control reserve not by an individual system but by an integrated energy and production system. Still, the schedule of the production system stays fixed if control reserve is requested. Only the energy system adapts the operation on request, but the production system schedule is optimized for the control-reserve provision by the integrated system. Thus, if control reserve is requested, the same amount of energy is demanded by the production system, but the operation of multiple energy conversion units is adapted and not only a single energy conversion unit increases or decreases the energy demand. To consider the changing operational cost, we model the operation of the energy system if negative or positive control reserve is requested explicitly. Therein, we assume that the entire amount of the offered balancing-power is requested.

First-stage decisions are for all time steps the amount of provided control-reserve capacity, the offered energy price, and the schedule of the production system (Figure 4.2). Second-stage decisions correspond to the actual operation of the energy system in each of the three scenarios for control-reserve requests (POS, NEG, NO) for all time steps. Both stages are coupled by the energy demand and purchased electricity. The energy demand of the production system is determined in the first stage, while the actual operation of the energy system to fulfill the energy demand depends on the request scenario. Furthermore, the amount of provided control-reserve capacity and energy price has to be decided before it is known if the control-reserve capacity is actually requested or not. Thus, the provided control-reserve capacity is considered by non-anticipativity constraints. Non-anticipativity constraints model that some decisions need to be taken at a certain time based only on the current information and cannot be changed once a scenario materializes (Higle, 2005). In Figure 4.2, we show the stochastic process for a single time step. As there is no time-coupling in the second-stage operation of the energy system, the scenario tree has a size of three second-stage scenarios per time step to model the request uncertainties.

In principle, another stage of uncertainty results from the bidding process for control-reserve capacity. Here, we neglect this uncertainty by assuming that the offered control-reserve capacity is always accepted. For this purpose, we employ average historical price data for the capacity prices. Under this assumption, acceptance is highly likely, but we can only gain average revenues for the control-reserve capacity.

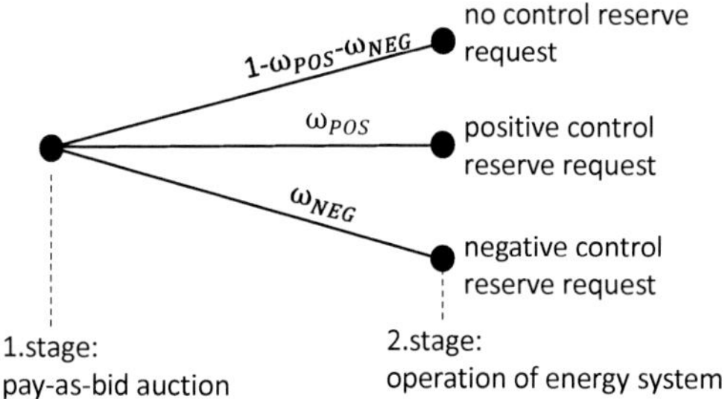

Figure 4.2: Stochastic process of the two-stage stochastic programming model for a single time step. On the 1. stage, the amount of provided control-reserve capacity and the energy price for delivered control reserve are decided in a pay-as-bid auction. In the 2. stage, the operation of the energy system needs to be decided, depending on whether positive, negative, or no control-reserve energy is requested. The respective scenarios are considered with their corresponding probabilities: ω_{NEG}, ω_{POS} and (1-ω_{POS}-ω_{NEG}).

4.1.2 Model extension for stochastic optimization

In this section, we present the model for the provision of control reserve by an integrated energy and batch production system. We present the changes of the model from the integrated scheduling without control-reserve provision. Thus, this model extends the model in Section 3.1, but the design variables are fixed. Our method determines the production schedule, the operation of the energy system if control reserve is requested or not requested, the offered control-reserve capacity, and the offered energy price. Inputs are the product demand of the production system, the gas price, the spot-market price for electricity, average capacity prices for control-reserve provision, and the probability of the control-reserve request. We formulate a stochastic programming model as MILP.

Control-reserve capacity is offered as indivisible. Offering indivisible control-reserve capacity is possible in some control-reserve markets which allows the provision of control reserve by market participants that only can supply a certain amount of control reserve. In this case, the control-reserve provider avoids the delivery of control reserve between the offered control-reserve capacity and zero control-reserve capacity. Thus, the model only needs to ensure the delivery of the offered control-reserve capacity. Furthermore, control-reserve capacity is only offered in integer values which is mandatory for some offers in control-reserve markets.

Objective Function

The model's objective is to minimize the cost for operating the energy system and the production system $C^{\mathrm{OPEX,total}}$. The objective function considers five cost terms:

$$\min C^{\mathrm{OPEX,total}} = \sum_{t \in T} (C_t^{\mathrm{OPEX,PS}} + C_t^{\mathrm{EL}} - R_t^{\mathrm{EL}} + C_t^{\mathrm{Gas,NO}} - R_t^{\mathrm{R,tot}}). \tag{4.1}$$

$C_t^{\mathrm{OPEX,PS}}$ describes the cost for operating the production system. C_t^{EL} is the cost of buying electricity from the grid. R_t^{EL} defines the revenue that is generated when selling electricity to the grid. $C_t^{\mathrm{Gas,NO}}$ describes the cost of buying gas from the grid if no control reserve is requested. The gas cost when control reserve is requested is considered in $R_t^{\mathrm{R,tot}}$, where $R_t^{\mathrm{R,tot}}$ describes the revenue for providing control reserve. The electricity cost is not changed if control reserve is requested. Thus, by request of control reserve, only the gas cost is changed. By minimizing these five cost terms, the cost for operating the overall system is minimized.

The constraints resulting from the stochastic programming compared to Chapter 3 are described in more detail in the following. Here, we provide the equations for the participation in the control-reserve market, which are the extension of the integrated scheduling problem.

Revenues of participating in the control-reserve market

In the objective (Equation (4.1)), the total revenues for providing control reserve $R_t^{\mathrm{R,tot}}$ are considered. The total revenues for providing control reserve $R_t^{\mathrm{R,tot}}$ are the sum of revenues from positive and negative control reserve. The revenues from positive ($cr = POS$) and negative ($cr = NEG$) control reserve $R_{t,cr,ep}^{\mathrm{R}}$ have three terms in every time step t, the first stage variable: revenues for providing control-reserve capacity, and the second stage variables: the extra cost for gas if control reserve is delivered and revenues from delivering control-reserve energy:

$$\begin{aligned} R_{t,cr,ep}^{\mathrm{R}} =& P_{t,cr}^{\mathrm{RP}} \cdot \mathrm{p}_{t,cr}^{\mathrm{CAP}} + \omega_{t,cr,ep} \cdot (C_t^{\mathrm{Gas,NO}} - C_{t,cr}^{\mathrm{Gas}}) \\ &+ P_{t,cr}^{\mathrm{RP}} \cdot \mathrm{p}_{t,cr,ep}^{\mathrm{R}} \cdot \omega_{t,cr,ep} \quad \forall t \in T, cr \in CR, ep \in EP_{cr}. \end{aligned} \tag{4.2}$$

The revenues for providing control-reserve capacity are determined by the amount of control-reserve capacity $P^{\text{RP}}_{t,cr}$, and the capacity price $\text{p}^{\text{CAP}}_{t,cr}$. These revenues are paid even if no control-reserve energy is requested and, thus, are not multiplied with the probability for control-reserve request. The extra cost for gas if control reserve is delivered is determined with the probability for control-reserve request $\omega_{t,cr,ep}$ and the difference between the gas cost if control reserve is requested $C^{\text{Gas}}_{t,cr}$ and if no control reserve is requested $C^{\text{Gas,NO}}_{t}$. The extra cost for gas are caused by different operation of the energy system to provide a different amount of electricity to the grid and can be negative.

The revenues from delivering control-reserve energy are determined by the amount of control-reserve capacity $P^{\text{RP}}_{t,cr}$, the energy price $\text{p}^{\text{R}}_{t,cr,ep}$, and the probability for control-reserve request $\omega_{t,cr,ep}$. In reality, the energy price is a continuous variable and the probability of request is a nonlinear function of the energy price: The higher the energy price, the lower the probability of request and vice versa. Furthermore, the energy price is multiplied by the offered control-reserve capacity, which is an integer variable. To obtain a MILP model, we linearize the last term in Equation (4.2). For this purpose, we discretize the energy price as parameter $\text{p}^{\text{R}}_{t,cr,ep}$, where the index *ep* distinguishes the discrete energy prices. Consequently, for each discrete energy price, we obtain the probability of request as parameter $\omega_{t,cr,ep}$. By the discretization of the energy price and probability of request, we limit our offers to discrete energy prices while we avoid to solve a MINLP.

At the control-reserve market, only a single energy price is offered for negative and positive control reserve each. This is contrary to some day-ahead electricity markets, where electricity supply can be offered dependent on the electricity price. Equation (4.3) ensures that only a single combination of positive energy prices ep^+ and negative energy prices ep^- is chosen. Thus, the binary variable λ_{t,ep^+,ep^-} equals 1 for the chosen energy price in every time step t:

$$\sum_{ep^+\in EP^+}\sum_{ep^-\in EP^-} \lambda_{t,ep^+,ep^-} = 1 \quad \forall t \in T. \tag{4.3}$$

The total revenue for providing control reserve $R^{\text{R,tot}}_t$ is then determined by:

$$R^{\text{R,tot}}_t = \sum_{ep^+\in EP^+}\sum_{ep^-\in EP^-} (R^{\text{R}}_{t,POS,ep^+} + R^{\text{R}}_{t,NEG,ep^-}) \cdot \lambda_{t,ep^+,ep^-} \quad \forall t \in T. \tag{4.4}$$

The multiplication of $R^{\text{R,tot}}_t$ and λ_{t,ep^+,ep^-} would result in a MINLP. We linearize Equation (4.4) by a Big-M reformulation:

$$R_t^{\mathrm{R,tot}} \leq R^{\mathrm{R}}_{t,POS,ep^+} + R^{\mathrm{R}}_{t,NEG,ep^-} + \mathrm{M}\cdot(1-\lambda_{t,ep^+,ep^-}) \quad \forall t \in T, ep^+ \in EP^+, ep^- \in EP^-. \tag{4.5}$$

Equation (4.5) considers the revenues for positive control reserve $R^{\mathrm{R}}_{t,POS,ep^+}$ and negative control reserve $R^{\mathrm{R}}_{t,NEG,ep^-}$. The binary variable λ_{t,ep^+,ep^-} chooses the energy price for positive and negative control reserve. M is a sufficiently large number. The revenues of participation in the control-reserve market $R_t^{\mathrm{R,tot}}$ are maximized by the objective function. Thus, Equation (4.4) can be replaced by the linearization in Equation (4.5).

In many control-reserve markets, control reserve has to be offered for predefined time slices. Hence, the amount of provided control-reserve capacity $P^{\mathrm{RP}}_{t,cr}$ needs to be offered for the duration of the time slice:

$$P^{\mathrm{RP}}_{t,cr} = P^{\mathrm{RP}}_{t+a,cr} \quad \forall t \in \{1, 1+\mathrm{t^s}, \ldots, 1+\mathrm{t^h}-\mathrm{t^s}\}, a \in \{1,2,\ldots,\mathrm{t^s}-1\}, cr \in CR. \tag{4.6}$$

$\mathrm{t^h}$ is the time horizon considered for scheduling the energy and production system. $\mathrm{t^s}$ is the length of each time slice. Also, the energy price needs to be offered for the duration of the time slice $\mathrm{t^s}$. Thus, the binary variable λ_{t,ep^+,ep^-} is equal:

$$\begin{aligned} \lambda_{t,ep^+,ep^-} = \lambda_{t+a,ep^+,ep^-} \; \forall t \in \{1, 1+\mathrm{t^s}, \ldots, 1+\mathrm{t^h}-\mathrm{t^s}\}, a \in \{1,2,\ldots,\mathrm{t^s}-1\}, \\ ep^+ \in EP^+, ep^- \in EP^-. \end{aligned} \tag{4.7}$$

Energy Balances

The energy balances couple the production system with the energy system, i.e., the energy demand of the production system has to be fulfilled by the energy system. To ensure that the energy system is able to provide the requested control reserve, the energy balances have to be defined for every scenario. Again, the three scenarios $cr \in \{NO,POS,NEG\}$ are distinguished. Equation (4.8) gives the energy balance for electricity for the three scenarios considering the electricity consumption on the left side and the electricity production on the right side:

$$E_{t,\text{el}}^{\text{demand}} + \zeta_{cr} \cdot P_{t,cr}^{\text{RP}} + \sum_{cu \in CU} U_{t,cu,\text{el},cr} + P_{t,\text{el}}^{\text{sell}} = P_{t,\text{el}}^{\text{buy}} + \sum_{pu \in PU} V_{t,pu,\text{el},cr} \qquad (4.8)$$
$$\forall t \in T, cr \in \{POS, NEG, NO\}.$$

The electricity balance considers the following variables: as first-stage variables: the electricity demand from the batch production system $E_{t,\text{el}}^{\text{demand}}$, the amount of control reserve provided $P_{t,cr}^{\text{RP}}$, the electricity $P_{t,\text{el}}^{\text{sell}}$ sold to the grid and the electricity purchased from the grid $P_{t,\text{el}}^{\text{buy}}$, and as second-stage variables: the power consumption $U_{t,cu,\text{el},cr}$ of the energy conversion units consuming electricity, e.g., compression chillers or electric boilers, and the electrical power generation $V_{t,pu,\text{el},cr}$ of the energy conversion units producing electricity, e.g., combined-heat-and-power engines. The parameter ζ_{cr} distinguishes the three scenarios and is 0 in the scenario 'no control reserve is requested' (NO), 1 in the 'scenario positive control reserve is requested' (POS), or (-1) in the scenario 'negative control reserve is requested' (NEG). The electricity $P_{t,\text{el}}^{\text{sell}}$ sold to the grid and the electricity purchased from the grid $P_{t,\text{el}}^{\text{buy}}$ do not depend on the scenario, since we assume that the exchange with the electricity market is already fixed when the participation in the control-reserve market is optimized. For the German day-ahead electricity market that is performed after the control-reserve market, the fixed exchange implies that the integrated system buys or sells the optimized amount of electricity no matter what electricity price materializes.

By the request of control reserve, the energy demand and, at the same time, the production schedule are not affected. Thus, we consider that only the energy system provides the request of control reserve. Furthermore, our method allows that control reserve could be requested in every time step. If control reserve is requested, the energy system operates a different set of energy conversion units to provide control reserve and supplies the same energy to the production system. Therefore, our method ensures security of energy supply no matter at which time and how long the offered control reserve is requested.

4.2 Case study

The proposed method is applied to the case study with the batch production system from Kondili et al. (1993) (Figure 3.2) and the energy system model from Baumgärtner et al. (2019a), which is an extension of the energy system model by Voll et al. (2013).

4.2.1 Description of the case study

The integrated energy and batch production system is assumed to participate in the German tertiary control-reserve market (50Hertz Transmission GmbH et al., 2019). In the German tertiary control-reserve market, offers can only be given as integer values if the offered control-reserve capacity is below 5 MW. Furthermore, offers can be declared as indivisible. Indivisible offers are only fully requested, where other offers can be requested partly. Here, we consider indivisible offers of control-reserve capacities between 0 MW and 5 MW. In Germany, tertiary control reserve is traded for 4 h time slices. If tertiary control reserve is requested, the control reserve needs to be provided within 15 min. We assume that the modeled energy conversion units fulfill this requirement.

In our case study, we model a day from 2018 for the capacity price and use historical data from the German tertiary control-reserve market. In our method, we discretized the energy price. From the historical data, we obtain for each time slice 4 energy prices for positive and negative control reserve and the respective probability of request. The probability of request is derived from historical data based on 50Hertz Transmission GmbH et al. (2019) and Bundesnetzagentur | SMARD.de (2020). For each considered energy price, we analyze how often and how long control-reserve capacity was requested in one year. The probability of request is then defined as the number of time steps in which control-reserve capacity was requested for the energy price divided by all analyzed time steps.

The electricity price for selling electricity is taken for each hour from the spot market price. The average selling price is 39 €/MWh, and for the average purchasing price, we increased this average selling price by 9.4 €/MWh.

For the batch production system, we added demands for low- and high-temperature heating, cooling and electricity for each task. We schedule the systems for a 24 h time horizon with a time step length of 1 h. The product demand to be fulfilled is 224 t of S07 and 432 t of S10. The equipment in the batch production system can process the following maximum batch size: E01 (80 t), E02 (90 t), E03 (70 t), E04 (70 t). The energy system has the following units and corresponding thermal capacity: 3 gas-driven boilers (2 MW, 1.5 MW, 0.5 MW), 3 electricity-driven boilers (2 MW, 1.5 MW, 0.5 MW), 3 compression chillers (2.5 MW, 1.5 MW, 0.5 MW), 3 absorption chillers (2.5 MW, 1.5 MW, 0.5 MW), 4 combined-heat-and-power engines (3 MW, 2.5 MW, 2 MW, 1.5 MW). Thus, the energy system consumes gas and can purchase or sell electricity to provide the production system with heat, cold, and electricity.

The optimization models are formulated in GAMS 27.3.0 (GAMS Development, 2020) and solved by ODH 4.2.6 with CPLEX 12.9.0.0 (IBM Corporation, 2020). The time limit is set to 7200 s with an optimality gap of 0.1 %. We compare our method with sequential scheduling with and without providing control reserve and integrated scheduling without providing control reserve.

4.2.2 Results

In the case study, the proposed method reaches cost savings of 4.6 % compared to the sequential scheduling without control reserve (Figure 4.3). Furthermore, the method saves 3.3 % compared to the sequential scheduling with control-reserve provision and 2.3 % compared to integrated scheduling without control reserve. The revenues from the control-reserve market in the proposed method are 3.1 %, which are larger than the cost savings compared to the integrated scheduling without control reserve. Thus, the method chooses a more expensive schedule for the integrated energy and production system, but by doing so, the method realizes substantial revenues in the control-reserve market.

In the proposed method, provision of control reserve is chosen in every time slice. But in one time slice, only negative control reserve is chosen. The chosen negative control-reserve capacity is the maximum amount of 5 MW from 0-20 h, and 1 MW from 20-24 h. The chosen positive control-reserve capacity is the maximum amount of 5 MW from 0-8 h, 3 MW from 8-12 h, 4 MW from 16-20 h and 1 MW from 20-24 h. From 12-16 h, no positive control reserve is chosen.

For negative and positive control reserve, different energy prices are chosen from the discretized energy prices. The chosen energy prices range from the lowest to the highest discretized energy prices. The different energy prices change the probability of request. Thus, by optimizing the energy price and the amount of control reserve provided, a trade-off is resolved: A higher energy price leads to a lower probability of request. The probability of request is not only a weighting factor for potential revenues from control reserve, but also weights the potential additional cost of changing the schedule of the energy system when control reserve is requested. Consequently, our method optimizes the schedule of both systems simultaneously such that the highest expected revenues are gained from the trade-off (Figure 4.3).

If we compare the cost savings by providing control reserve isolated for sequential and integrated scheduling, participating in the control-reserve market is more beneficial in the integrated scheduling than in the sequential scheduling: Cost savings in the integrated scheduling are 2.3 % compared to 1.3 % in the sequential scheduling.

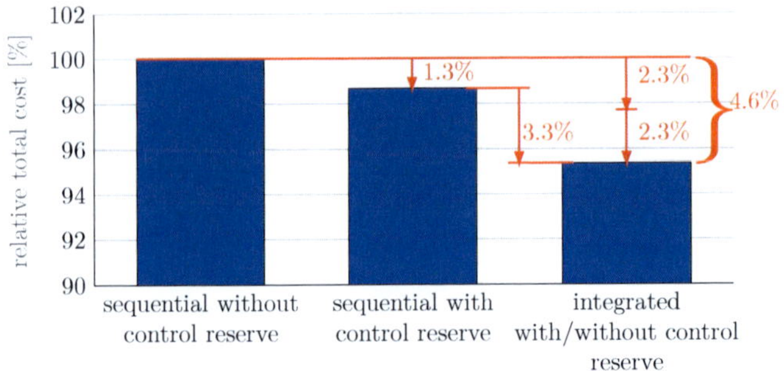

Figure 4.3: Relative total cost for energy and production system for sequential scheduling and integrated scheduling. Both approaches are applied with and without the possibility to provide control reserve.

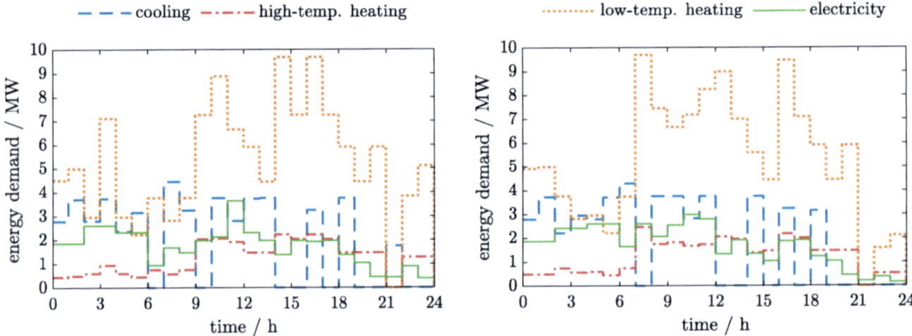

Figure 4.4: Energy demand of the production system for the integrated scheduling without (left) and with (right) provision of control reserve. temp.=temperature.

The sequential scheduling is solved to optimality with consideration of control reserve in 995 s wall time and without consideration of control reserve in 966 s wall time. In sequential optimization with and without consideration of control reserve, the optimization of the energy system is fast (below 40 s), and the scheduling of the production system takes 960 s. In the integrated scheduling without providing control reserve, a gap of 0.5 % is reached within the time limit. In our method for the integrated scheduling with provision of control reserve, a gap of 4.9 % remains. We also performed the integrated scheduling with an increased time limit of 24 h time limit, but the gap remained high at 3.7 %. Nevertheless, our method finds a solution with lower cost compared to the best benchmark already within 188 s.

The energy demands are different in the integrated scheduling without and with the provision of control reserve (Figure 4.4). As stated in Section 4.1, the energy demand of the production system remains unchanged if control reserve is requested. The peaks in the energy demand do not differ significantly between the integrated scheduling without and with the provision of control reserve. Thus, the energy demand is only shifted. This shifting of the energy demand is most significant in the demand of low-temperature heating.

The amount of energy provided by the different types of energy conversion units differs between the scheduling methods. If control reserve is provided in the integrated scheduling, the combined-heat-and-power units provide on average only 69 % of the heat compared to 78 % if no provision of control reserve is considered (Figure 4.5). The gas-driven boilers increase their share of provided heat from 22 % to 28 % if control reserve is provided. A single electricity-driven boiler is operated in only 1 time step if no control reserve is provided. The electricity-driven boiler provides 3 % of the heat demand if control reserve is provided. Thus, a wider variety of operated energy conversion units enables the provision of control reserve.

If positive control reserve is requested, the combined-heat-and-power units increase their share of heat supply from 69 % to 82 % (Figure 4.5 (lower left)). Additionally, the gas-driven boilers decrease their share of heat supply, and the electricity-driven boilers are idle. If negative control reserve is requested, the combined-heat-and-power units decrease their share of heat supply to 33 % (Figure 4.5 (lower right)). At the same time, the gas-driven boilers and the electricity-driven boilers increase their share of heat supply to 40 % and 27 %, respectively.

From 21 h to 22 h, the production system has no heat demand but requires cooling and electricity in integrated scheduling with and without provision of control reserve. The energy system provides the cooling demand only with compression chillers if no control reserve is requested. If positive control reserve is requested, the cooling demand is partly supplied by an absorption chiller. The heat to run the absorption chiller is then supplied by a combined-heat-and-power engine, which also provides the requested positive control reserve. In the same way, if negative control reserve is requested, the absorption chiller is supplied by an electricity-driven boiler, which then supplies the negative control reserve. From 12 h to 16 h, no positive control reserve is provided. Thus, the supply structure of the energy system is the same as if no control reserve is requested in these time steps (Figure 4.5).

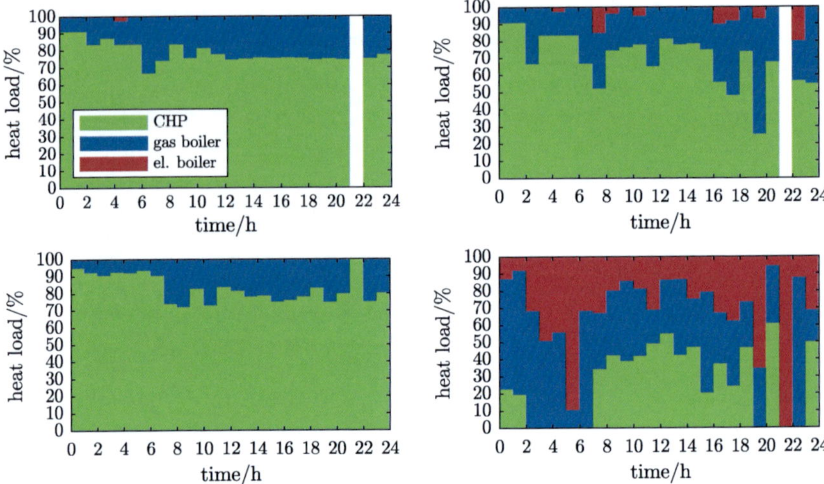

Figure 4.5: Share of heat supply by the energy conversion units for the integrated scheduling without (upper left) and with (upper right) provision of control reserve if no control reserve is requested. Furthermore, the share of heat supply by the energy conversion units is given if only positive control reserve (lower left) and only negative control reserve (lower right) is requested. The actual operation is a mix of (upper right), (lower left) and (lower right), depending on the actual control reserve requested. El.=Electricity-driven.

Concluding, control reserve is provided by changing the operation of all energy conversion units. Furthermore, the optimal operation of the energy system is even different without and with provision of control reserve if no control reserve is requested.

4.3 Conclusions

Control reserve is commonly provided either by an energy system or a production system. However, if both systems are located in the same industrial site, the integrated system can provide control reserve more efficiently. By integrated scheduling, the benefits of the provision of control reserve can be increased.

In this chapter, we propose a method for integrated scheduling of energy and batch production system for control-reserve provision. The production schedule and, consequently, its energy demand is optimized for provision of control reserve by the energy system. The method optimizes the offered amount of negative and positive control reserve in control-reserve markets. We consider the request probability and the expected revenues from providing the requested control reserve in a stochastic programming model. Furthermore, the schedule of the energy system is provided if control reserve is requested. The method considers an average capacity price and, thus, no uncertainty for acceptance of the offered capacity price.

The method is applied to a case study and shows large cost savings compared to sequential scheduling with control-reserve provision (3.3 %) and integrated scheduling without control-reserve provision (2.3 %). The optimization did not reach the optimality gap in the given time limit, but the best solution is found rapidly and already outperforms the other approaches. Thus, the method enables energy and production system operators to schedule integrated systems for participation in control-reserve markets.

In this part, we showed that the application of integrated optimization models for energy and production systems is superior compared to sequential optimization. However, if both systems do not share the same objective function, an integrated optimization cannot be applied. For this purpose, in the following Part II, we present methods to solve a Stackelberg game between energy and production systems.

Part II

Stackelberg-based optimization

Chapter 5

Scheduling coordination of energy and production system

The integrated scheduling of energy and production systems is not applicable if the objectives of the systems are misaligned. If the objectives of energy and production systems are misaligned, we describe the situation as a Stackelberg game (c.f. Section 2.3). Consequently, we assume that energy and production system are operated by different companies (Figure 2.1). In a Stackelberg game, the leader can either have complete or incomplete information on the follower (c.f. Section 2.2). In this chapter, we present a method to schedule energy and production systems in a Stackelberg game with incomplete information. The production system is the leader announcing the energy demand, and the energy system is the follower announcing the energy cost. In the method, we approximate the demand-dependent energy cost of the energy system and provide this as incomplete information to the production system.

The chapter is organized as follows: In Section 5.1, the proposed method for solving the Stackelberg game with incomplete information is presented. Therein, we provide a detailed description of the determination of the demand-dependent energy cost. In Section 5.2, the method is applied to two case studies. In Section 5.3, conclusions are presented.

Major parts of this chapter are reproduced by permission of Elsevier with modifications from the complete Elsevier source from:

Leenders, L., Bahl, B., Hennen, M., and Bardow, A. (2019a). Coordinating scheduling of production and utility system using a Stackelberg game. *Energy*, 175, 1283-1295.

The author of this thesis contributed to the development and implementation of the method as well as the calculation and interpretation of the results, wrote the draft of the paper and is its principal author.

5.1 Minimizing total costs in a Stackelberg game

In this section, we propose a method to minimize the total costs $C^{\text{min TOTAL}}$ to be paid by the production system. The proposed method is based on a repeated Stackelberg game (Section 5.1.1). In the repeated Stackelberg game, first, the production system minimizes the production costs C^{PS}, while considering estimated energy costs $C^{\text{ES,est}}$. The estimated energy costs are calculated based on demand-dependent energy costs, which are introduced in Section 5.1.2. Secondly, the energy system optimizes its operation to fulfill the energy demand $E^{\text{demand}}_{t,e}$ of the production system. In Section 5.1.3, the models of energy and production system are provided.

5.1.1 Repeated Stackelberg game

The proposed method to minimize the total costs models a Stackelberg game, where the production system is the leader, and the energy system is the follower. The Stackelberg game is repeated until a convergence criterion is satisfied. The repeated Stackelberg game (Figure 5.1) consists of the following steps:

1. Initially, the production system assumes constant prices as demand-dependent energy costs. The counter for the repetitions of the Stackelberg games is initialized ($i = 1$), and the total costs of a virtual previous loop are set to infinity ($C^{\text{TOTAL}}_{i=0} = \infty$) for the convergence criterion (Step 5).

2. The production system determines an optimal production schedule with minimal costs $C^{\text{TOTAL,est}}_{i}$ considering costs for both energy and production. The energy costs $C^{\text{ES,est}}$ are estimated based on the demand-dependent energy cost (Section 5.1.2). The energy demand $E^{\text{demand}}_{t,e}$ resulting from the production schedule is transferred to the energy system.

3. The energy system minimizes its operational costs $C^{\text{ES,cost}}_{i}$ to supply the energy required for the production schedule. The energy system can then add a profit margin to the operational costs ($C^{\text{ES,offer}}_{i} \geq C^{\text{ES,cost}}_{i}$). The energy costs $C^{\text{ES,offer}}_{i}$ are transferred back to the production system.

4. The production system calculates the total costs C^{TOTAL}_{i} as a sum of the production costs C^{PS}_{i} and the offered energy costs $C^{\text{ES,offer}}_{i}$, i.e., the offered energy costs from the energy system replace the estimated energy costs based on the demand-dependent energy costs.

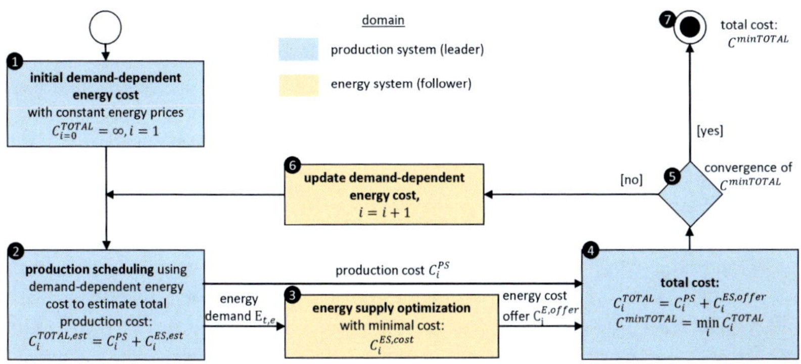

Figure 5.1: Method for minimizing total costs by repeated Stackelberg game between energy and production system.

5. The production system checks the convergence criterion: Have the minimal total costs $C^{\text{min TOTAL}}$ improved within a specified number of previous repetitions of the Stackelberg game?

6. If the convergence criterion is not satisfied, the energy system updates the demand-dependent energy costs (Section 5.1.2) for the production system. The production system minimizes the total costs using the updated demand-dependent energy costs such that the Stackelberg game is repeated (Steps 2-5).

7. If the convergence criterion is satisfied, the Stackelberg game is stopped by the production system: The method has identified the minimal total costs $C^{\text{min TOTAL}}$ based on incomplete information exchange.

In the proposed method, the Stackelberg game is repetitively solved until the convergence criterion is met. However, the production system can only approximate the energy system behavior using incomplete information, i.e., demand-dependent energy cost. Thus, the identification of the optimum is not guaranteed. The proposed repeated Stackelberg game allows minimizing the operational cost of the production system. In particular, the proposed repeated Stackelberg game provides only incomplete information on the cost structure of the energy system to the production system. Thus, the method reflects the situation often encountered in practice when energy and production system are run by different companies.

5.1.2 Demand-dependent energy costs

The demand-dependent energy costs (Step 6) calculated by the energy system are an important part of the proposed method. The demand-dependent energy costs reflect the cost of energy supply by the energy system without revealing all information on the energy system to the production system. The demand-dependent energy costs provide energy prices in 4 energy price bands for every energy form e (Figure 5.2): two energy price band around the current energy demand $E_{t,e}^{\text{curr}}$ of energy form e, one band for increasing energy demand, and one band for decreasing energy demand. By the energy price bands, the energy system can reflect the change of its operational costs with changing the energy demand. To model this complex relationship between costs and energy demand practically, we determine the operational costs for 4 additional energy demands. The cost structure is then linearly interpolated between these operation points and extrapolated (Figure 5.2). Suitable energy demands are identified based on the assumption that the price for the energy supply changes significantly when a running energy conversion unit can be switched off, or an additional energy conversion unit needs to be switched on. The energy demand when an energy conversion unit is switched on ($E_{t,e}^{\text{lb1}}$) or off ($E_{t,e}^{\text{ub1}}$) is then used to define the energy price bands:

- Energy price band for large decrease:

$$E_{t,e}^{\text{min}} \leq E_{t,e} < E_{t,e}^{\text{lb1}} \quad \forall t \in T, e \in E. \tag{5.1}$$

- Energy price band for small decrease:

$$E_{t,e}^{\text{lb1}} \leq E_{t,e} \leq E_{t,e}^{\text{curr}} \quad \forall t \in T, e \in E. \tag{5.2}$$

- Energy price band for small increase:

$$E_{t,e}^{\text{curr}} \leq E_{t,e} \leq E_{t,e}^{\text{ub1}} \quad \forall t \in T, e \in E. \tag{5.3}$$

- Energy price band for large increase::

$$E_{t,e}^{\text{ub1}} < E_{t,e} \leq E_{e}^{\text{max}} \quad \forall t \in T, e \in E. \tag{5.4}$$

To calculate the bounds $E_{t,e}^{\text{lb1}}$ and $E_{t,e}^{\text{ub1}}$, the set of all energy conversion units in the energy system ($u \in U$) is divided into 2 subsets ($U = U_t^{\text{on}} \dot{\cup} U_t^{\text{off}}$): the running energy conversion units (U_t^{on}) and the idle units (U_t^{off}) for every time step t. The energy price band for small increase and decrease cover the range where no additional energy

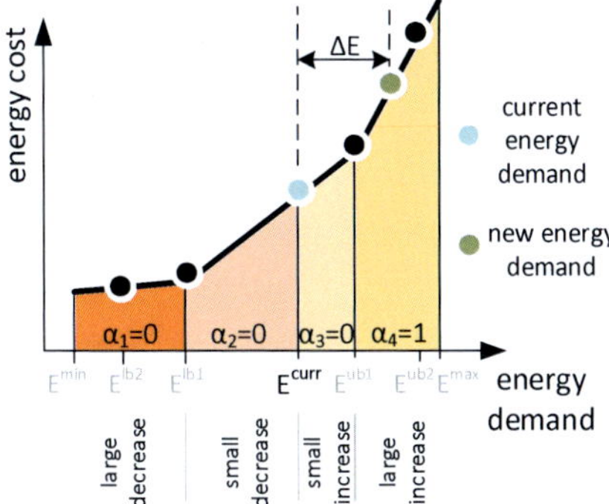

Figure 5.2: The energy price bands in the demand-dependent energy cost. The binary variable α_s indicates in which linear section the energy demand $E_{i,e}^{\text{curr}} + \Delta E_{i,e}$ is located. This figure is reproduced from Leenders et al. (2021) by permission of Elsevier with modifications from the complete Elsevier source.

conversion unit needs to be switched on or off to satisfy the energy demand. At the upper bound $E_{t,e}^{\text{ub1}}$ of the current energy price band, all running energy conversion units operate at their possible full load $V_{u,e}^{\text{N}}$:

$$E_{t,e}^{\text{ub1}} = \sum_{u \in U_t^{\text{on}}} V_{u,e}^{\text{N}} \quad \forall t \in T, e \in E. \tag{5.5}$$

At the lower bound $E_{t,e}^{\text{lb1}}$ of the current energy price band, the smallest running energy conversion unit can just not be switched off:

$$E_{t,e}^{\text{lb1}} = \left(\sum_{u \in U_t^{\text{on}}} V_{u,e}^{\text{N}} \right) - \left(\min_{u \in U_t^{\text{on}}} V_{u,e}^{\text{N}} \right) + \varepsilon \quad \forall t \in T, e \in E, \tag{5.6}$$

with ε being a small power value, e.g., $\varepsilon = 1\,\text{kW}$.

Any change $\Delta E_{t,e}$ of the current energy demand $E_{t,e}^{\text{curr}}$ changes the energy costs. To pass information on this dependence between energy demand and costs from the energy system to the production system, we introduce the demand-dependent energy cost (Figure 5.2). We approximate the change in energy costs with the energy demand by linear interpolation between the energy costs at the upper and lower bound of the price band. Thereby, the costs change by $c^{\mathrm{k}}_{t,e} \cdot \Delta E_{t,e}$ due to a change in the energy demand of size $\Delta E_{t,e}$ is obtained.

The demand-dependent energy cost for small decrease $c^{\text{curr-}}_{t,e}$ is determined by the energy system for every energy form e (Figure 5.2):

$$c^{\text{curr-}}_{t,e} = \frac{C^{\mathrm{ES}}(E^{\text{curr}}_{t,e}) - C^{\mathrm{ES}}(E^{\text{lb1}}_{t,e})}{(E^{\text{curr}}_{t,e} - E^{\text{lb1}}_{t,e}) \cdot \Delta t} \quad \forall t \in T, e \in E. \tag{5.7}$$

The energy costs of the energy demand is $C^{\mathrm{ES}}(E_{t,e})$.

The demand-dependent energy cost for small increase $c^{\text{curr+}}_{t,e}$ is determined by the energy system for every energy form e (Figure 5.2):

$$c^{\text{curr+}}_{t,e} = \frac{C^{\mathrm{ES}}(E^{\text{ub1}}_{t,e}) - C^{\mathrm{ES}}(E^{\text{curr}}_{t,e})}{(E^{\text{ub1}}_{t,e} - E^{\text{curr}}_{t,e}) \cdot \Delta t} \quad \forall t \in T, e \in E. \tag{5.8}$$

If the energy demand changes beyond the range of the small decrease or small increase, energy conversion units are turned on or off such that the price structure changes for the energy system. This change of the price structure is reflected by the demand-dependent energy costs $c^{\text{upper}}_{t,e}$ and $c^{\text{lower}}_{t,e}$ for the large increase and large decrease energy price band, respectively. These demand-dependent energy costs are calculated similarly as for the energy price bands in Equation (5.7) and Equation (5.8), i.e.,

$$c^{\text{upper}}_{t,e} = \frac{C^{\mathrm{ES}}(E^{\text{ub2}}_{t,e}) - C^{\mathrm{ES}}(E^{\text{ub1}}_{t,e})}{(E^{\text{ub2}}_{t,e} - E^{\text{ub1}}_{t,e}) \cdot \Delta t} \quad \forall t \in T, e \in E, \tag{5.9}$$

$$c^{\text{lower}}_{t,e} = \frac{C^{\mathrm{ES}}(E^{\text{lb1}}_{t,e}) - C^{\mathrm{ES}}(E^{\text{lb2}}_{t,e})}{(E^{\text{lb1}}_{t,e} - E^{\text{lb2}}_{t,e}) \cdot \Delta t} \quad \forall t \in T, e \in E. \tag{5.10}$$

In Equation (5.9) and Equation (5.10), two further energy demands are introduced to capture the impact of changing energy demands on the price structure in the other price bands. The demand $E^{\text{ub2}}_{t,e}$ corresponds to the situation when all running energy conversion units (U^{on}_t) are operating at full load plus the smallest energy conversion unit that was idle is also operating at full load:

$$E^{\text{ub2}}_{t,e} = \sum_{u \in U^{\text{on}}_t} V^N_{u,e} + \min_{u \in U^{\text{off}}_t} V^N_{u,e} \quad \forall t \in T, e \in E. \tag{5.11}$$

In contrast, the demand $E^{\mathrm{lb2}}_{t,e}$ reflects the situation where the second smallest running energy conversion unit can just not be switched off:

$$E^{\mathrm{lb2}}_{t,e} = (\sum_{u \in U^{\mathrm{on}}_t} V^{\mathrm{N}}_{u,e}) - \min_{u,m \in U^{\mathrm{on}}_t, u \neq m} (V^{\mathrm{N}}_{u,e} + V^{\mathrm{N}}_{m,e}) + \varepsilon \quad \forall t \in T, e \in E. \tag{5.12}$$

The demand-dependent energy costs are calculated for every time step t. Thus, we solve $4 \cdot |T|$ optimization problems to calculate the demand-dependent energy cost.

The energy costs of the current energy demand $C^{\mathrm{ES}}(E^{\mathrm{curr}}_{t,e})$ and the demand-dependent energy costs $c^{k}_{t,e}$ are transferred from the energy system to the production system together with the bounds $E^{\mathrm{ub1}}_{t,e}$ and $E^{\mathrm{lb1}}_{t,e}$. If desired, a profit margin can be added by the energy system. The proposed demand-dependent energy costs condense the complex cost structure of the energy system to inform the production system. In the next repetition of the Stackelberg game (Figure 5.1), the production system employs the updated demand-dependent energy costs $c^{\mathrm{k}}_{t,e}$ to estimate the energy costs for its production schedule. These estimated energy costs ($C^{\mathrm{ES,est}}$) are considered in the objective function of the production system. In the next section, the system modeling is described in detail.

5.1.3 System modeling

In this section, we provide the main equations of the energy and production systems optimization models in addition to the given equations in Chapter 3. Each system is considered by a separated optimization problem formulated as a MILP model. The information exchanged between both systems is the energy demand, transferred from the production system to the energy system, and the demand-dependent energy costs, transferred vice versa (Figure 5.1). In the energy system model, no time-coupling constraints are considered, e.g., no storage units are considered. If time-coupling constraints are considered in the energy system model, the calculation of the demand-dependent energy costs needs to be adapted.

5.1.3.1 Production system

The production system operates in batch mode and is again modeled by the State-Task-Network formulation (Kondili et al., 1993; Shah et al., 1993) (Chapter 3). The product demand is fixed and has to be fulfilled at the end of the time horizon $t \in T$. The estimated total costs $C^{\mathrm{TOTAL,est}}$ are the objective of the production system.

$C^{\text{TOTAL,est}}$ consists of two costs parts: the production costs C^{PS} and the estimated energy costs $C^{\text{ES,est}}$:

$$C^{\text{TOTAL,est}} = C^{\text{PS}} + C^{\text{ES,est}}. \tag{5.13}$$

The production costs C^{PS} considers costs for operating the production equipment $j \in J$ in time steps $t \in T$ for performing task $i \in I$:

$$C^{\text{PS}} = \sum_{t\in T}\sum_{j\in J}\sum_{i\in I}\sum_{t'=t-\Delta t_{i,j}^{\text{process}}+1}^{t} (B_{t',i,j} \cdot OC_{i,j}^{\text{var}} + W_{t',i,j} \cdot OC_{i,j}^{\text{fix}}). \tag{5.14}$$

The production costs C^{PS} consists of two parts: The first part is linearly depending on the batch size $B_{t,i,j}$ with operation costs $OC_{i,j}^{\text{var}}$ and the second part represents the fixed operating costs $OC_{i,j}^{\text{fix}}$ when a unit is running as indicated by the binary variable $W_{t',i,j}$.

The estimated energy costs $C^{\text{ES,est}}$ are calculated from the energy costs of the previous repetition of the Stackelberg game $C_{t,e}^{\text{ES}}$ plus any changes $\Delta E_{t,e}^{k}$:

$$C^{\text{ES,est}} = \sum_{e\in E}\sum_{t\in T}(C_{t,e}^{\text{ES}} + \sum_{k\in\{upper,curr+,curr-,lower\}} c_{t,e}^{\text{k}} \cdot \Delta E_{t,e}^{\text{k}}). \tag{5.15}$$

The changes of the energy demand $\Delta E_{t,e}^{\text{upper}}$, $\Delta E_{t,e}^{\text{curr+}}$, $\Delta E_{t,e}^{\text{curr-}}$ and $\Delta E_{t,e}^{\text{lower}}$ in each price band have different specific costs given by the corresponding demand-dependent energy costs $c_{t,e}^{\text{upper}}$, $c_{t,e}^{\text{curr+}}$, $c_{t,e}^{\text{curr-}}$ and $c_{t,e}^{\text{lower}}$ (Figure 5.2). The changes of the energy demand $\Delta E_{t,e}^{\text{upper}}$, $\Delta E_{t,e}^{\text{curr+}}$, $\Delta E_{t,e}^{\text{curr-}}$ and $\Delta E_{t,e}^{\text{lower}}$ are used to calculate the energy demand $E_{t,e}^{\text{new}}$ in the current repetition of the Stackelberg game:

$$E_{t,e}^{\text{new}} = E_{t,e}^{\text{curr}} + \sum_{k\in\{upper,curr+,curr-,lower\}} \Delta E_{t,e}^{\text{k}} \quad \forall t \in T, e \in E. \tag{5.16}$$

The exact mathematical modeling of $E_{t,e}^{\text{new}}$ by $\Delta E_{t,e}^{\text{upper}}$, $\Delta E_{t,e}^{\text{curr+}}$, $\Delta E_{t,e}^{\text{curr-}}$ and $\Delta E_{t,e}^{\text{lower}}$ in Equation (5.16) requires several binary variables and linearization of bilinear terms, which is presented in Appendix A.2. Finally, the energy demand $E_{t,e}^{\text{new}}$ is passed to the energy system as updated $E_{t,e}^{\text{curr}}$ (Figure 5.1).

5.1.3.2 Energy system

The energy system is modeled according to the MILP formulation from Chapter 3 (Voll et al., 2013). The objective of the energy system is to minimize the costs for energy supply:

$$C^{\text{ES,cost}} = \sum_{t \in T} [\Delta t \cdot (P^{\text{buy}}_{t,el} \cdot p^{\text{buy}}_{el} - P^{\text{sell}}_{t,el} \cdot p^{\text{sell}}_{el} + U^{\text{gas}}_{t} \cdot p^{\text{buy}}_{gas})]. \tag{5.17}$$

The costs for energy supply $C^{\text{ES,cost}}$ of the energy system considers costs for:

- Buying electricity $P^{\text{buy}}_{t,el}$ from the grid at a price of p^{buy}_{el},
- Selling electricity $P^{\text{sell}}_{t,el}$ to the grid at a price of p^{sell}_{el},
- Buying gas U^{gas}_{t} from the grid at a price of p^{buy}_{gas}.

The energy demand $E^{\text{demand}}_{t,e}$ of the production system has to be satisfied by the output energy $E^{\text{supply}}_{t,e}$ of the energy system:

$$E^{\text{demand}}_{t,e} = E^{\text{supply}}_{t,e} \quad \forall t \in T, e \in E. \tag{5.18}$$

All other equations of the energy system are identical to Voll et al. (2013) and are given in Appendix A.1.2. In the modeling, we consider part-load efficiencies and minimal part-load.

5.2 Case studies

The proposed repeated Stackelberg game is applied to two scheduling case studies from literature (Kallrath (2002) (Figure 3.9) and Kondili et al. (1993) (Figure 3.2)). The case studies are again extended by energy systems. In the case studies, the Stackelberg game ends if there is no improvement in the total costs $C^{\text{min TOTAL}}$ compared to the last repetition of the Stackelberg game (Step 5, Figure 5.1). The initial constant energy prices are set to zero to indicate the effectiveness of the proposed method, even if no information is available on the energy price. In this case study, the energy price bands for a small increase and decrease use a joint energy price calculated between $E^{\text{ub1}}_{t,e}$ and $E^{\text{lb1}}_{t,e}$. This assumption is extended in Chapter 6.

The repeated Stackelberg game is compared to a sequential optimization, where the energy system provides no feedback to the production system. As target value, we also minimize the total costs by an integrated optimization of energy and production system (Part I). The integrated optimization defines the minimal total costs achievable and is thus a lower bound for the repeated Stackelberg game.

To show the effectiveness of the proposed demand-dependent energy costs (Section 5.1.2), we execute the same repeated Stackelberg game without demand-dependent energy costs transferred from the energy system. Instead, the energy system transfers only the energy price to the production system. Furthermore, we compare with the case that the energy system transfers not only the energy price for the current energy demand but also the energy price derivative which is the energy price in the energy price band for small decrease/increase to the production system.

The optimization problems of energy and production system (Section 5.1.3) are formulated as a MILP in GAMS 24.7.3 (GAMS Development, 2020) and solved with CPLEX 12.6.3.0 (IBM Corporation, 2020). All optimization problems are solved with an optimality gap of 0.5 %. In Section 5.2.1 and Section 5.2.2, the case studies are described, and in Section 5.2.3, the computational results are discussed.

5.2.1 Case study I

The production system proposed by Kallrath (2002) has 9 production equipment performing 17 different tasks (Figure 3.9 and Table 5.1). The product demand is fixed (State 16: 24 t; State 17: 24 t; State 19: 24 t) and has to be fulfilled at the end of the production time (33 h with $\Delta t = 1$ h). We enhance the case study by introducing heating and electricity demands to each task. Additional parameters are provided in Appendix B.3. The assumed energy system consists of 3 energy conversion units (1 boiler with 3000 kW, 1 boiler with 1000 kW thermal capacity, and 1 CHP engine with a thermal capacity of 3000 kW). The energy system model is based on Voll et al. (2013).

5.2.2 Case study II

The production system proposed by Kondili et al. (1993) has 4 production equipment performing 5 different tasks (Figure 3.2 and Table 5.2). The product demand is fixed (State 7: 117 t; State 10: 216 t) and has to be fulfilled at the end of the production time (30 h with $\Delta t = 1$ h). We also enhanced this case study by introducing heating and electricity demands to each task. Additional parameters are provided in Appendix B.4. The assumed energy system consists of 5 energy conversion units (3 boilers with 500 kW, 1500 kW, and 4000 kW thermal capacity, and 2 CHP engines with a thermal capacity of 1500 kW each). Again, the energy system model is based on Voll et al. (2013).

Table 5.1: Equipment and scheduling data of the production system for case study I. E: production equipment; T: task.

equipment	suitable task (task duration /h)	form of energy demand	minimum and maximum batch size /t
E1	T1(2)	electricity	3-10
E2	T2(4)	heat	5-20
E3	T3(2)	heat	4-10
E4	T4(4); T5(4); T6(4); T7(4)	heat & electricity	4-10
E5	T8(6); T9(6)	electricity	4-10
E6	T10(4); T11(5); T12(6)	electricity	3-7
E7	T10(5); T11(6); T12(6)	electricity	3-7
E8	T13(6); T14(4); T15(4);T16(6); T17(6)	heat & electricity heat & electricity electricity	4-12
E9	T13(6); T14(6); T16(6); T17(6)	heat & electricity heat & electricity electricity	4-12

Table 5.2: Equipment and scheduling data of the production system for case study II. E: production equipment; T: task.

equipment	suitable task (task duration /h)	form of energy demand	minimum and maximum batch size /t
E1	T1(1)	heat	10-80
E2	T2(2); T3(2); T4(1)	heat & electricity	30-90
E3	T2(2); T3(2); T4(1)	heat & electricity	70
E4	T5(2)	heat & electricity	40-100

5.2.3 Results

We apply the proposed repeated Stackelberg game (Section 5.1) to the case studies (Section 5.2.1 and Section 5.2.2). In both case studies, the proposed repeated Stackelberg game ends after the first repetition because the total costs increase in the second repetition (Figure 5.3).

The total costs from the repeated Stackelberg game are compared to the sequential optimization approach (=Repetition 0). In case study I, the total costs decrease by 9.1 %, and in case study II, the total costs decrease by 4.8 % (Figure 5.3, blue crosses). The total costs decrease by integrated optimization is 9.2 % in case study I and 5.6 % in case study II. Thus, in case study I, the proposed repeated Stackelberg game exploits nearly the complete optimization potential between sequential optimization and integrated optimization. In case study II, 80 % of the optimization potential is exploited.

However, the proposed Stackelberg game is a heuristic and cannot guarantee to reach the integrated optimum by further repetitions. For illustration, further repetitions of the Stackelberg game are calculated (Figure 5.3). In both case studies, total costs of the Stackelberg game vary only slightly (case study I: 0.01 %, case study II: 0.37 %).

For comparison, we further execute the repeated Stackelberg game without demand-dependent energy costs. The following price information is used for these repeated Stackelberg games (Figure 5.3):

- the energy price and the derivative of the energy price, which corresponds to the demand-dependent energy cost of the energy price bands around the current energy demand in each time step (Section 5.1.2), and
- only the energy price for the current energy demand (fixed energy price) in each time step.

In both case studies, the Stackelberg games without demand-dependent energy costs result in higher total costs than the proposed method using the demand-dependent energy costs.

The Stackelberg game using only the energy price derivatives also reduces the total costs significantly, but the lowest costs are still higher than for the proposed demand-dependent energy costs (case study I: 1.7 %; case study II: 0.1 %).

In both case studies, the Stackelberg game with only the fixed energy price also reduces the cost compared to the sequential approach but yields worse results than

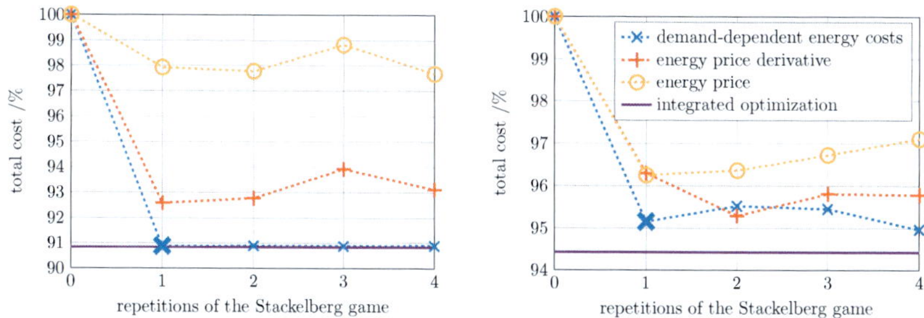

Figure 5.3: Total costs compared to the sequential approach (=100 %) for the proposed repeated Stackelberg game with the demand-dependent energy costs for case study I (left) and case study II (right). Furthermore, results are shown for the integrated optimization as well as the Stackelberg game using only the energy price and the energy derivative. The proposed Stackelberg game with demand-dependent energy costs ends after the first repetition, further repetitions are only shown for illustration.

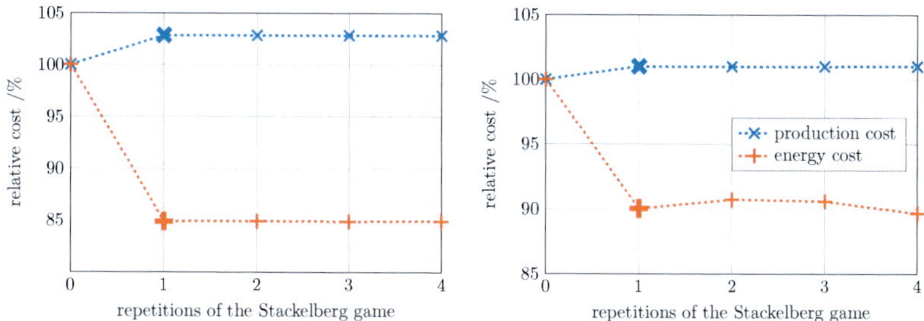

Figure 5.4: Relative costs for production (C_i^{PS}) and energy ($C_i^{\mathrm{ES,cost}}$) compared to the sequential approach for case study I (left) and case study II (right).

the Stackelberg game using the energy price derivatives. Thus, using less information as feedback to the production system leads to worse results in the Stackelberg game.

Analyzing the results of the repeated Stackelberg game in more detail allows to identify the origin of the cost savings (Figure 5.4). In both case studies, the proposed repeated Stackelberg game saves costs in the energy system, whereas the costs in the production system increase. The cost savings in the energy system result from different energy demand profiles (Figure 5.5 and Figure C.1). The energy supply

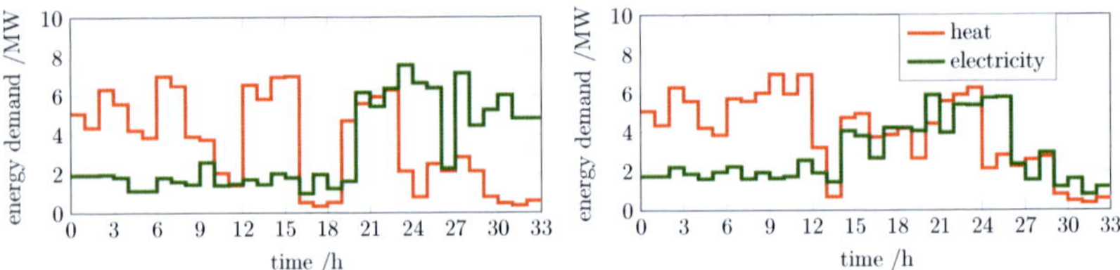

Figure 5.5: Energy demand of the sequential approach (left) and the proposed method (right) for case study I.

structure changes, e.g., in case study I, 51.9 % of the electricity demand is supplied by the grid in the sequential approach. In the Stackelberg game, only 23.1 % of the electricity demand is supplied by the grid. Consequently, the CHP engine supplies more electricity as well as more heat. The heat provided by the CHP engine rises from 57.0 % to 61.8 %. Thus, in case study I, in the Stackelberg game, more energy is supplied by the CHP engine. Still, the overall energy demand of the production system is nearly the same. Thus, the boiler is used less. For both case studies, the proposed repeated Stackelberg game has also been executed with initial energy prices greater than zero. The results are provided in Appendix C.2, showing that the proposed method leads to significant cost reductions independent from initial energy prices.

We also solved both case studies with the General-Column-Generation (GCG) solver (Gamrath and Lübbecke, 2010; Gleixner et al., 2018) that solves a Dantzig-Wolfe decomposition by Branch-Cut-and-Price. The GCG solver allows for both automated or user-specified decompositions. The integrated problems are decomposed into independent subproblems and a connective master problem. Here, we apply the user-specified decomposition mode. Our subproblems are the optimization problems of the energy system and of the production system. This decomposition is motivated by practice where the two separated optimization problems of energy and production system are solved. The master problem contains the equation connecting the energy demand and the energy supply as well as the integrated objective function.

For case study I, the resulting optimization problems are solved on an Intel(R) Core(TM) i7-8700 CPU @ 3.20GHz with 32GB RAM with a time limit of 100 h. No feasible solution is found in case study I. The optimization problem reaches the working memory of 32GB after 66h and thus is aborted.

For case study II, the resulting optimization problems are solved on a Dell Optiplex 990, i7-2600 @ 3.40GHz with 16GB RAM with a time limit of 100 h. For case study II, the solution reduces the cost compared to the sequential approach by 1.1 %. Thus, the Dantzig-Wolfe decomposition provides better solutions than the sequential approach. However, after the time limit is exceeded, there is still a gap of 3.7 % to the results from the proposed repeated Stackelberg game.

The solution provided by the GCG solver might improve by using the automated decomposition. However, the automated decomposition might not provide subproblems suitable for the practically motivated subproblems of production system and energy system. Thus, the results indicate the advantage of our proposed solution method, even compared with state-of-the-art decomposition methods.

In both case studies, the proposed method decreases the overall cost significantly in just one repetition. Based on these results, one repetition seems promising also for future applications. One repetition requires only one rescheduling of the plant operation. Thus, a real-world application of the proposed method results in moderate extra effort while significant cost savings can be expected. In summary, the proposed repeated Stackelberg game with demand-dependent energy costs leads to significant cost savings with only one repetition compared to the common sequential optimization approach.

5.3 Conclusions

In this chapter, a method is proposed to minimize the energy and production costs for the production system using incomplete information. The proposed method is based on a repeated Stackelberg game. In the repeated Stackelberg game, the production system is the leader, and the energy system is the follower. Demand-dependent energy costs are introduced to pass incomplete information on the costs of energy supply without revealing all information on the energy system to the production system.

In two presented case studies, the repeated Stackelberg game leads to 9.1 % and 4.8 % savings. These savings realize nearly 100 % and more than 80 % of the maximal possible savings from an integrated optimization of energy and production system. We compare the demand-dependent energy costs to other incomplete information with less detail, such as only the energy cost for the current energy demand. In the comparison, the demand-dependent energy costs reach the lowest cost. Thus, the proposed repeated Stackelberg game with the demand-dependent energy costs efficiently uses incomplete information between energy and production system to reach production schedules with lower total costs.

The method proposed in this chapter coordinates between a single energy system and a single production system using demand-dependent energy costs as incomplete information. However, in some industrial sites, multiple energy and production systems are operated. Because each system is optimized individually in the proposed method, the method can be extended by additional coordination methods among the energy systems and among the production systems. This extension of the method is proposed in the following Chapter 6.

Chapter 6

Scheduling coordination of multiple energy and production systems

In the previous chapter, we presented a method to coordinate a single batch production and a single energy system. The systems have misaligned objectives, and thus, we described the situation as a Stackelberg game. Consequently, we assumed that energy and production system are operated by different companies (Figure 2.1). The method considers incomplete information exchange between systems based on demand-dependent energy cost. In this chapter, we extend the method to multiple energy and production systems. For this purpose, we propose coordination methods. In particular, we propose a method to allocate demand-dependent energy cost to the production systems. Furthermore, the energy systems compete in minimizing the energy price. For this purpose, we present an optimization problem for energy price minimization.

The chapter is organized as follows: In Section 6.1, the extended method for solving the multi-leader multi-follower Stackelberg game with incomplete information is presented. We describe the three coordination methods, i.e., coordination between energy systems and production systems, coordination among the energy systems and coordination among the production systems. In Section 6.2, the method is applied to two case studies. In the first case study, two production systems and two energy systems are coordinated. In the second case study, four production systems and two energy systems are coordinated. In Section 6.3, conclusions are presented.

Major parts of this chapter are reproduced by permission of Elsevier with modifications from the complete Elsevier source from:

Leenders, L., Ganz, K., Bahl, B., Hennen, M., Baumgärtner, N., and Bardow, A. (2021). Scheduling coordination of multiple production and utility systems in a multi-leader multi-follower Stackelberg game. *Computers & Chemical Engineering*, 150, 107321.

The author of this thesis contributed to the development and implementation of the method as well as the calculation and interpretation of the results, wrote the draft of the paper and is its principal author.

6.1 Coordination of multiple energy and production systems

We consider the following setup of the Stackelberg game (Figure 6.1): multiple production systems are operated on an industrial site. The production systems are supplied with energy by multiple energy systems. The energy supply is organized in 4 steps:

1. The production systems (leader) announce their energy demands based on fixed energy prices to the energy systems (follower), e.g., hourly energy demands for one day.
2. The energy systems (follower) respond by announcing the associated hourly energy cost. The energy cost depends on the operation of the energy systems to supply the energy demand.
3. The production systems (leader) reschedule their operation with the knowledge of their energy cost, also leading to a rescheduled energy demand.
4. The energy systems (follower) adapt their operation to the rescheduled energy demand and announce the associated hourly energy cost.

We propose a method that uses only incomplete information to coordinate multiple energy and production systems in a multi-leader multi-follower Stackelberg game. Still, with only incomplete information available, our method reduces cost significantly.

In our method, all systems optimize themselves. Thus, each production system and each energy system schedules itself. The systems are coordinated based on incomplete information corresponding to 1) energy demands of the production systems and 2) demand-dependent energy cost announced by the energy systems. The coordination can be performed by an authority or any participating production or energy system. For an even higher degree of confidentiality, an aggregator can anonymize the data and pass it to a coordinator as Wenzel et al. (2020) propose.

The proposed method for coordination has to solve 3 problems (Figure 6.1):

- Problem a): coordination between energy systems and production systems (Section 6.1.1)
  - How to exchange incomplete information for potential demand-side management?
- Problem b): coordination of the energy systems (Section 6.1.2)
  - How to identify the amount of energy supplied by each energy system?

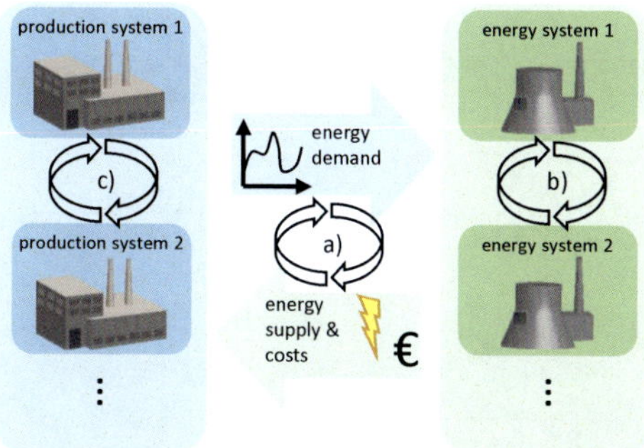

Figure 6.1: Problem setup for coordination between energy and production systems illustrated for two production systems and energy systems. The production systems announce their energy demand. The energy demand is allocated to the energy systems. The energy systems announce the energy cost for the supply of the energy demand. The three problems to be solved are: a) How to coordinate between energy systems and production systems? (Section 6.1.1) b) How to coordinate the energy systems? (Section 6.1.2) c) How to coordinate the production systems? (Section 6.1.3).

  - How to condense individual demand-dependent energy cost to overall demand-dependent energy cost?

- Problem c): coordination of the production systems (Section 6.1.3)
  - How to allocate the demand-dependent energy cost to the individual production systems?
  - How do changes in the energy demand of the individual production systems affect the overall demand-dependent energy cost?

6.1.1 Coordination between production systems and energy systems

The operation of energy and production systems needs to be coordinated because each system's operation is affected by the operation of the other systems. The production systems schedule their production for minimal cost. Their cost is caused by running the production processes and from the energy cost. In industry, energy cost is often not simply proportional to the energy demand because the energy costs depend on the operation of the energy systems. The proposed method coordinates among all

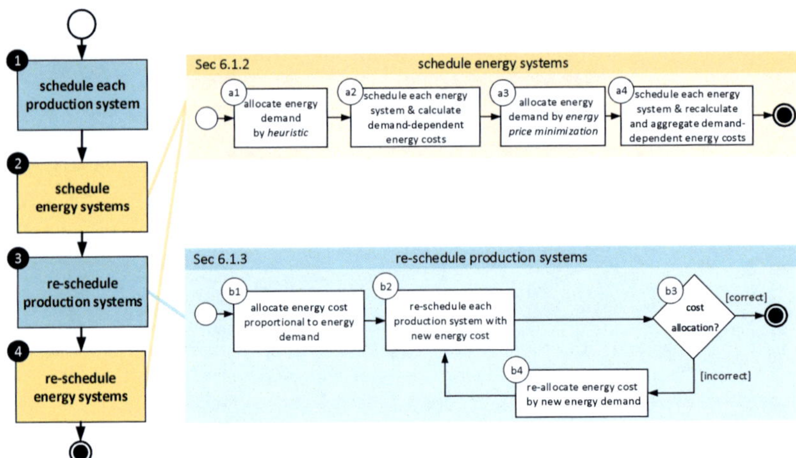

Figure 6.2: Method to solve the multi-leader multi-follower Stackelberg game by coordination between multiple energy and production systems. Two inner algorithms are employed to schedule energy systems (Step ②+④) and production systems (Step ③).

systems, while only incomplete information is exchanged. As in Chapter 5, we use the concept of demand-dependent energy cost.

The coordination of multiple energy and production systems is performed in 4 steps (left-hand side of Figure 6.2):

- Step ①: The production systems schedule their production. For the cost of energy, constant prices are assumed. Step ① determines the energy demand as input for Step ②.
- Step ②: The energy systems schedule their operation (details in Section 6.1.2). In Step ②, the energy systems coordinate how much energy is supplied by each energy system. The coordination aims for a minimal energy price. Furthermore, each energy system calculates its demand-dependent energy cost. Afterward, the demand-dependent energy costs of all energy systems are aggregated. This aggregated demand-dependent energy cost is the input for Step ③.
- Step ③: The production systems re-schedule their production (details in Section 6.1.3). The energy cost of each production system is identified. For this purpose, aggregated demand-dependent energy costs from Step ② and the en-

ergy demands from the re-scheduling are used as incomplete information. The rescheduled energy demand of the production systems is passed to the energy systems as the input for Step ④.

- Step ④: The energy systems reschedule their operation to identify the final energy cost (details in Section 6.1.2). As in Step ②, the energy systems coordinate how much energy is supplied by each energy system. The result of Step ④ is the energy cost to be paid by the production systems.

The steps ②, ③, and ④ use inner algorithms to coordinate the systems. The inner algorithms are explained in the following Sections 6.1.2 and 6.1.3. A crucial element of the method is the demand-dependent energy cost that is determined by each energy system. The concept of demand-dependent energy cost is described in detail in Section 5.1.2.

In the following section, we present the coordination between the systems.

6.1.2 Coordination among energy systems: Step ② + ④

The energy demand of the production systems is fulfilled by multiple energy systems. The energy systems compete for the amount of energy that each energy system supplies to the production systems. The coordination has to identify how much energy is supplied by each energy system. We assume that the coordination among the energy systems aims for the minimum energy cost. Thus, the coordination allocates the energy demand to reach the minimal cost of energy supply.

For each energy system, the cost of energy supply does not have one fixed value but depends on the energy demand. This dependence needs to be captured by the coordination method. For this purpose, we use the demand-dependent energy cost.

To reach the minimal energy price, the coordination between the energy systems is performed in 4 substeps (Figure 6.2; a1-a4):

- Substep a1: The energy demand is allocated to the energy systems. In Step ②, the allocation is proportional to the maximum capacity of each energy system. In Step ④, the allocation is based on the amount of energy provided by each energy system, as calculated in Step ②.

- Substep a2: Each energy system schedules itself to fulfill the allocated energy demand from Substep a1. Furthermore, each energy system calculates its demand-dependent energy cost.

- Substep a3: The coordinating system solves the optimization problem to minimize the energy price. The optimization problem reallocates the energy demand. As a result, the amount of energy provided by each energy system is determined. The details of the optimization problem are given below.
- Substep a4: As in Substep a2, each energy system is scheduled, but with the reallocated energy demand from Substep a3. Based on the new schedule, each energy system recalculates its demand-dependent energy cost. Subsequently, all demand-dependent energy costs are aggregated to one aggregated demand-dependent energy cost, such that the lowest energy cost is reached. This aggregated demand-dependent energy cost is a merit-order curve for the energy supply.

In Step ② of the main method, the output of the coordination is the aggregated demand-dependent energy cost and the energy cost of the current energy demand. In Step ④, the output of the coordination is the energy cost to fulfill the energy demand.

Energy price minimization in Substep a3

The coordination between the energy systems allocates how much energy is supplied by each energy system. This allocation of the energy demand uses demand-dependent energy cost from Substep a2 to represent the energy cost of each energy system. Thus, only incomplete information is exchanged. The objective of the allocation is to minimize the energy price.

The energy demand is allocated by an optimization problem. The optimization problem is performed separately for every energy form e, e.g., heat, electricity. In the following, the equations of the optimization problem are given. The objective is to minimize the energy price c_e:

$$\min \quad c_e. \tag{6.1}$$

The main constraints are the energy balances. In the previous Substep a2, each energy system calculates its demand-dependent energy cost for the current energy demand. Thus, the current energy demand is already fulfilled by the energy systems. Consequently, if one energy system supplies additional energy, the remaining energy systems have to supply less energy by the same amount. Thus, the sum of differences in provided energy $\Delta E_{i,e}$ by all energy systems equals zero:

$$\sum_i \Delta E_{i,e} = 0. \tag{6.2}$$

$\Delta E_{i,e}$ is the difference of energy provided by energy system i compared to Substep a2.

The energy cost of each energy system is considered by its demand-dependent energy cost (Figure 5.2). Since the demand-dependent energy cost only reflect the cost change if the energy demand in one energy form is changed, an energy demand change in multiple energy forms might worsen the approximation of the energy cost by the demand-dependent energy cost. The modeling of the demand-dependent energy cost is given in Section 5.1.2.

The objective of the optimization problem is the minimization of the overall energy price c_e to fulfill the energy demand (Equation (6.2)). Thus, the energy price of each energy system i needs to be lower or equal to the overall energy price c_e. In the following equations, we calculate the energy price of each energy system i. We use the difference in the energy supply $\Delta E_{i,e}$, the difference between the energy demands to define the demand-dependent energy cost and the current energy demand $\Delta E^{k}_{i,e}$, and the cost of the energy demands $C^{\mathrm{k}}_{i,e}$. We rearrange the equations to avoid a possible division by an energy demand of 0:

large decrease:

$$c_e \cdot (\Delta E_{i,e} + E^{\mathrm{curr}}_{i,e}) \geq \alpha_{i,e,1} \cdot \left[\frac{C^{\mathrm{lb2}}_{i,e} - C^{\mathrm{lb1}}_{i,e}}{\Delta E^{\mathrm{lb2}}_{i,e} - \Delta E^{\mathrm{lb1}}_{i,e}} \cdot (\Delta E_{i,e} - \Delta E^{\mathrm{lb2}}_{i,e}) + C^{\mathrm{lb2}}_{i,e} \right] \quad \forall i \in I \tag{6.3}$$

small decrease:

$$c_e \cdot (\Delta E_{i,e} + E^{\mathrm{curr}}_{i,e}) \geq \alpha_{i,e,2} \cdot \left[\frac{C^{\mathrm{lb1}}_{i,e} - C^{\mathrm{curr}}_{i,e}}{\Delta E^{\mathrm{lb1}}_{i,e}} \cdot \Delta E_{i,e} + C^{\mathrm{curr}}_{i,e} \right] \quad \forall i \in I \tag{6.4}$$

small increase:

$$c_e \cdot (\Delta E_{i,e} + E^{\mathrm{curr}}_{i,e}) \geq \alpha_{i,e,3} \cdot \left[\frac{C^{\mathrm{ub1}}_{i,e} - C^{\mathrm{curr}}_{i,e}}{\Delta E^{\mathrm{ub1}}_{i,e}} \cdot \Delta E_{i,e} + C^{\mathrm{curr}}_{i,e} \right] \quad \forall i \in I \tag{6.5}$$

large increase:

$$c_e \cdot (\Delta E_{i,e} + E^{\mathrm{curr}}_{i,e}) \geq \alpha_{i,e,4} \cdot \left[\frac{C^{\mathrm{ub2}}_{i,e} - C^{\mathrm{ub1}}_{i,e}}{\Delta E^{\mathrm{ub2}}_{i,e} - \Delta E^{\mathrm{ub1}}_{i,e}} \cdot (\Delta E_{i,e} - \Delta E^{\mathrm{ub1}}_{i,e}) + C^{\mathrm{ub1}}_{i,e} \right] \quad \forall i \in I. \tag{6.6}$$

The energy price of each energy system i is determined with the energy cost from the demand-dependent energy cost [square brackets] divided by energy supplied by each energy system (round brackets). The binary variable $\alpha_{i,e,s}$ indicates in which linear section the energy demand $E_{i,e}^{\text{curr}} + \Delta E_{i,e}$ is located (Figure 5.2).

The energy price is defined to be greater 0:

$$c_e \geq 0. \tag{6.7}$$

$C_{i,e}^{\sim}$ is the energy cost for the energy demand $(\Delta E_{i,e}^{\sim} + E_{i,e}^{\text{curr}})$. Because the two continuous variables energy price c_e and difference of energy demand $\Delta E_{i,e}$ are multiplied, the optimization problem is a mixed-integer nonlinear program (MINLP).

The result of the energy-price minimization is the allocation of how much energy is supplied by each energy system $(\Delta E_{i,e} + E_{i,e}^{\text{curr}})$. Based on the allocation, the demand-dependent energy cost is recalculated and aggregated in Substep a4. The aggregated demand-dependent energy cost is the input of Step ③. Step ③ is explained in more detail in the following section.

6.1.3 Coordination among production systems: Step ③

In Step ①, the production systems schedule their production assuming fixed energy prices. For the resulting energy demand, in Step ②, the energy systems optimize their operation and determine the energy cost for the given energy demand. In Step ③, the production systems reschedule their production. As a price signal, the energy systems provide aggregated demand-dependent energy costs. The aggregated demand-dependent energy cost approximates the energy cost for changing energy demands. Here, the costs of all energy systems are aggregated into a single cost curve. This aggregated demand-dependent energy cost is considered in Step ③ to reschedule the production systems.

In Step ③, each production system reschedules the production independently. For this purpose, each production system needs to know its demand-dependent energy cost. Thus, we need to allocate the aggregated demand-dependent energy cost to the individual production systems. For this purpose, we perform substeps in Step ③ such that each production system assumes the correct energy cost while rescheduling only its own production.

This coordination requires 4 Substeps b1-b4 (Figure 6.2):

- Substep b1: The energy costs from Step ② are allocated according to the current energy demand of each production system. Thus, the specific energy price for

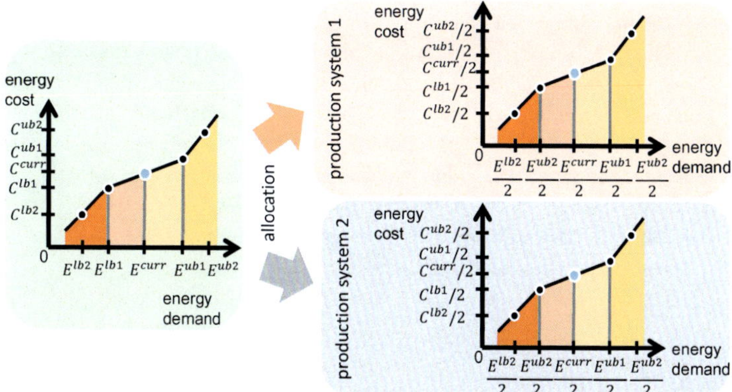

Figure 6.3: Allocation of aggregated demand-dependent energy cost to two production systems with equal energy demands. On the left-hand side, the aggregated demand-dependent energy cost is shown. The energy cost is allocated proportionally to the energy demand. Since both production systems have the same energy demand, the aggregated demand-dependent energy cost is split in half (right-hand side). Here, we illustrate the allocation for an aggregated demand-dependent energy cost with only 4 linear sections and, thus, one energy system. Generally, the aggregated demand-dependent energy cost has 4 times as many linear sections as energy systems are present.

the current energy demand is the same for each production system, but the energy cost differs due to the different current energy demands. Consequently, each production system has an individual base of the demand-dependent energy cost. The linear sections of the aggregated demand-dependent energy cost, e.g., from $E^{\text{curr}}_{i,e}$ to $(E^{\text{curr}}_{i,e} + \Delta E^{\text{lb1}}_{i,e})$, are equally allocated to all production systems (Figure 6.3). Thus, all production systems have linear sections of the same absolute size but with different current energy demands.

- Substep b2: Each production system reschedules the production considering its allocated demand-dependent energy cost from Substep b1.

- Substep b3: The coordination is finished if no production system changes its energy demand compared to the previous iteration or the maximum number of iterations is reached. Otherwise, Substep b4 is performed.

- Substep b4: Substep b4 removes potential errors from the allocation used in Substep b2. In Substep b2, every production system reschedules independently and, consequently, the energy demands have changed independently. Each production system calculates the energy cost based on its allocated demand-dependent energy cost from Substep b1. However, the sum of energy cost calculated by each production system individually might not be consistent with the energy cost from the non-allocated demand-dependent energy cost. To achieve consistent cost, the allocation from Substep b1 is revised in Substep b4. Details on the reallocation are explained at the end of this section.

The output of Step ③ is the energy demand of all production systems.

Reallocation of demand-dependent energy cost (Substep b4)

In Substep b2, the production systems reschedule their production based on the demand-dependent energy cost. In the rescheduling, each production system calculated its energy cost independently. However, these energy costs might not be the same as obtained by summing all energy demands and using the aggregated demand-dependent energy cost. This inconsistency is fixed in Substep b4.

The potential inconsistency is illustrated in Figure 6.4. In this example, the energy demand of production system 1 does not change by the rescheduling, while the energy demand of production system 2 increases strongly. The rescheduled energy demand of production system 2 is located in the linear section for large increases in energy demand. Thus, production system 2 assumes to pay a high cost for its energy demand. However, suppose the energy demands of production system 1 and 2 are added, and the energy costs are determined with the non-allocated demand-dependent energy cost. In that case, the actual energy cost of production system 2 is lower. The energy cost is lower because the energy demand is now correctly located in the linear section for a small increase.

We fix this inconsistency by reallocating the demand-dependent energy cost as follows: we determine the amount of the linear section for small increases that is not used by production system 1 with the rescheduled energy demand. The available energy supply with the lower energy price is distributed equally among the other production systems where the rescheduled energy demand increased strongly. The redistribution is also illustrated in Figure 6.4. By this redistribution, the energy cost calculated independently by each production system equals the energy cost from the non-allocated demand-dependent energy cost. The same principle for reallocation is used for decreasing the energy demand.

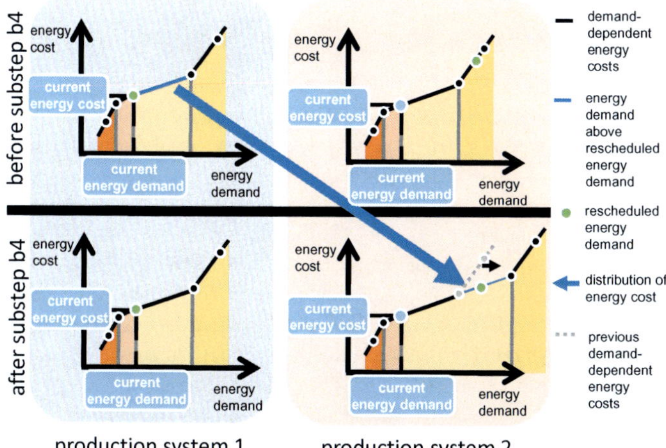

Figure 6.4: Example for the reallocation of the demand-dependent energy cost for two production systems (Substep b4). In the upper part of the figure, the demand-dependent energy cost of both production systems is shown. Both demand-dependent energy costs are equal. The energy demand of production system 1 (left) is not changed. The energy demand of production system 2 (right) is increased strongly and exceeds the section for small increases in the energy demand. Thus, in Substep b4, the linear section for small increases is extended for production system 2 by the amount not used by production system 1.

6.2 Case studies

The proposed method to coordinate the scheduling of multiple energy and production systems is applied to 2 case studies. In case study I (Section 6.2.1), we model 2 production systems and 2 energy systems. In case study II (Section 6.2.2), we apply our method to a larger case study by modeling 4 production systems and 2 energy systems. Furthermore, we generate 10 instances of both case studies. We apply our method to the instances and, thereby, show the benefits of the proposed method more generally. Our method is compared to the sequential and the integrated optimization. The sequential optimization represents the benchmark approach. A second benchmark is the integrated optimization that employs the total cost for the operation of the production system and the energy supply. Integrated optimization represents a best case that would require the exchange of complete information and the control of the production systems over the operational decisions of the energy systems.

In both case studies, we propose that the energy systems pass their energy cost on to the production systems without an additional profit margin. Thus, the overall cost is the sum of production cost and energy cost. The sequential optimization identifies minimal production cost but with potential higher energy cost. The proposed methods might increase production costs while reducing energy costs such that the overall cost is decreased. This decrease is beneficial for the production systems, while there are currently no incentives for the energy system in the given setting to decrease the energy cost. Thus, in reality, the production systems have to share their profit from the proposed method with the energy systems. Alternatively, the energy system might not share all cost reduction with the production system. In the case studies, we do not propose a method to share profits since we want to present the changes in energy and production costs without the impact of a concept for profit sharing. However, as overall cost decreases while supplying the same amount of product, profit sharing should be mutually beneficial for all parties.

6.2.1 Case study I

6.2.1.1 Description

In this case study, we model an industrial site with 2 production systems and 2 energy systems. The production systems are based on the case studies from Kallrath (2002) (Figure 3.9) and Kondili et al. (1993) (Figure 3.2). For each task of the production systems, we added electricity and heat demands. The product demands of the production systems are fixed at the time horizon of 30 h. Production system 1 (Kallrath, 2002) has the following product demand: 20 t of State 16; 20 t of State 17, and 20 t of State 19 (Figure 3.9). Production system 2 (Kondili et al., 1993) has the following product demand: 300 t of State 7 and 550 t of State 10 (Figure 3.2).

Both energy systems are based on the model by Voll et al. (2013). Energy system 1 has 2 boilers (3 MW, 1 MW) and 1 combined-heat-and-power engine (3 MW) for the energy supply. Energy system 2 has 3 boilers (4 MW, 1.5 MW, 0.5 MW), and 2 combined-heat-and-power engines (each 1.5 MW) for the energy supply. The energy systems can buy electricity from the grid for 0.16 €/kWh, sell electricity for 0.1 €/kWh and buy gas for 0.06 €/kWh. For the sequential optimization, each production system is supplied by only one energy system. Energy system 1 supplies production system 1, and energy system 2 supplies production system 2. A sequential optimization with an integrated optimization of the production systems and an integrated optimization of the energy systems would be slightly beneficial but is not considered here.

All optimization problems are formulated in GAMS 32.1.0 (GAMS Development, 2020). The scheduling problems of the energy and production systems (MILP) are solved with CPLEX 12.10.0.0 (IBM Corporation, 2020). The time limit to schedule the production systems is set to 7200 s, and the optimality gap is set to 0.5 %. The maximum number of iterations in Step ③ is set to 10. The time limit of the integrated optimization is set to 86,400 s. For the instances, the time limit of all optimization problems is set to 5000 s. The scheduling of the energy systems is solved within a few seconds to optimality. The optimization problem for the energy price minimization (MINLP) is solved with DICOPT (Kocis and Grossmann, 1989) using CONOPT 3.17L (Drud, 1996) for solving the NLPs and CPLEX 12.10.0.0 for solving the MILPs.

6.2.1.2 Results

In case study I, the proposed method saves 8.5 % of total production cost compared to the sequential optimization (Figure 6.5). The integrated optimization saves 9.7 % compared to the sequential optimization (Figure 6.5). Thus, the proposed method reaches 88 % of the potential cost reduction by an integrated optimization.

The proposed method decreases total costs by decreasing energy costs. The energy cost is 16.6 % lower than in the sequential optimization, while the production cost increases by 3.1 % (Figure 6.5). In sum, the overall cost decreases.

The reduction in energy cost results from different energy demands in the proposed method and the sequential optimization (Figure 6.6): The proposed method decreases the peak demand for electricity by 24 % compared to the sequential optimization. The peak of the heat demand increases by 13 %. Furthermore, the overall electricity demand decreases by 16.7 %, while the heat demand increases by 7 %. In absolute numbers, the sum of heat and electricity demand only changes slightly (−2 %). The energy demand is shifted from electricity to heat by choosing different tasks to produce the desired products. The changed electricity demand results in a lower share of electricity from the electricity grid: In the sequential optimization, 32.5 % of the electricity is supplied by the electricity grid compared to only 5.1 % in the proposed method. The integrated optimization even reduces the electricity supply from the electricity grid to 0.4 %.

The proposed method proceeds as follows: In Step ①, the production systems are scheduled while assuming fixed energy prices. In this case study, we assume the grid

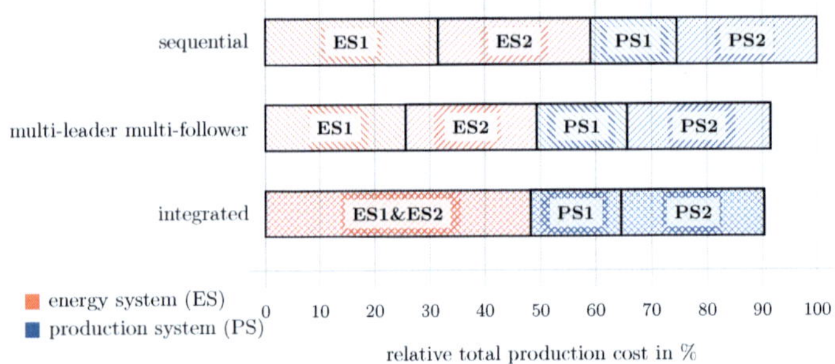

Figure 6.5: Cost in case study I for the different optimization approaches: the common sequential optimization between each production system and the corresponding energy system (sequential = 100 %), the multi-leader multi-follower Stackelberg game solved by the proposed coordination method (multi-leader multi-follower) and the integrated optimization of all systems (integrated). In the integrated optimization, the cost of the energy systems cannot be assigned to the different systems because the cost for additional electricity cannot be allocated unambiguously.

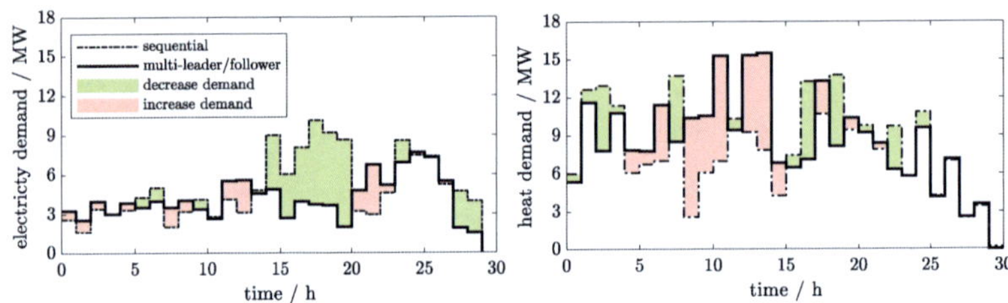

Figure 6.6: Case study I: energy demand of the production systems for the sequential optimization and for the proposed method to coordinate the multi-leader multi-follower Stackelberg game (electricity (left) and heat (right) demand).

price for electricity and the gas price for heat. In Step ②, the energy systems optimize their operation for the energy demand from Step ①. In Substep a2 of Step ②, the energy demand is heuristically allocated to the energy systems. Optimization leads to overall cost that is already 3.3 % lower than in the sequential optimization. In the following Substep a3, the energy demand is allocated based on an energy price minimization. For the newly allocated energy demand, the energy systems reoptimize

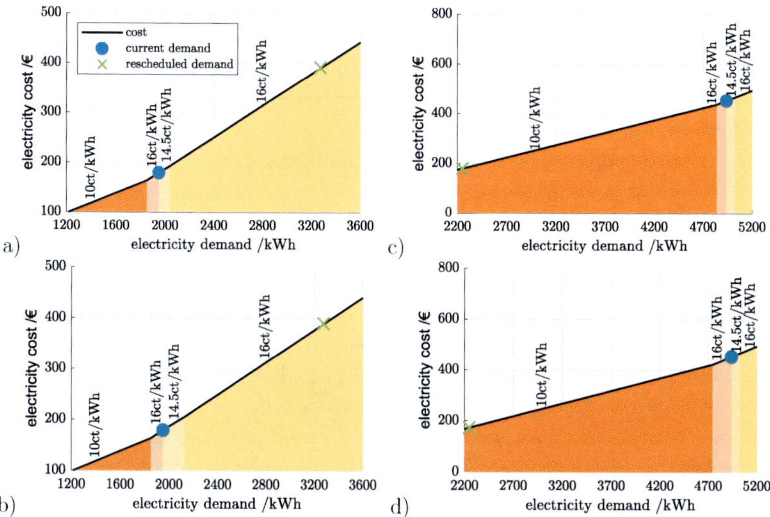

Figure 6.7: Case study I: demand-dependent electricity cost in time step 12 during iteration 1 (a,c) and 2 (b,d) of Step ③. Demand-dependent electricity costs of production system 1 are shown in (a) and (b) and demand-dependent electricity costs of production system 2 are shown in (c) and (d). From iteration 1 to iteration 2, production system 1 increases the electricity demand such that the electricity demand extends the section for *small increases*. Thus, the linear section for *small increases* is increased for production system 1 since production system 2 decreases its electricity demand. By this extension of the linear section for *small increases*, the electricity cost for production system 1 decreases. For production system 2, the linear section for *small decreases* is extended similarly since the electricity demand decreases such that the electricity demand extends the section for *small decreases*.

their operation in Substep a4, leading to overall cost savings of 3.2 % compared to the sequential optimization. Thus, the savings of Step ② are already obtained in Substep a2 because the heuristic allocation of the energy demand works already well in this case study. Here, the heuristic allocation is based on the maximum capacity of the energy systems. For applications to other energy systems, the energy-price minimization might have a larger impact because in this case study, we use the same component models and thus similar efficiency curves for the components in the energy systems.

If the energy systems have different components, the savings from the energy-price minimization in Substep a3 are expected to increase.

The results of Step ② are the demand-dependent energy costs in each time step. The demand-dependent energy costs reflect the prices for heat and electricity paid from the production systems to the energy systems. For each time step, we obtain demand-dependent energy costs.

In Step ③, the production systems are rescheduled in 4 iterations. Within these iterations, the demand-dependent energy cost is reallocated until the production systems do not change their energy demand anymore. We exemplary show the reallocation of the demand-dependent energy cost of production systems 1 and 2 from iteration 1 to iteration 2 in time step 12 for electricity (Figure 6.7). From iteration 1 to iteration 2, production system 1 increases the electricity demand by 1325 kW h, while production system 2 decreases the electricity demand by 2690 kW h. The increase of production system 1 is larger than the range for small increase (87 kW h) in iteration 1 and thus corresponds to a large increase. The decrease of production system 2 is larger than the range for small decrease (98 kW h) in iteration 1 and thus corresponds to a large decrease (cf. Figure 5.2). In iteration 2, the cost ranges are therefore reallocated. In particular, the range for small increases is extended for production system 1 and the range for small decreases is extended for production system 2. By this extension of the linear sections for small increase/decrease, the electricity cost decreases for the production systems.

In Figure 6.7, we also present the specific demand-dependent electricity cost. For small decreases in electricity, the specific cost is 0.16 €/kWh. For large decreases in electricity, the specific cost is 0.1 €/kWh, which equals the price of selling electricity to the grid. For small increases in electricity, the specific cost is 0.145 €/kWh and for large increases in electricity, the specific cost is 0.16 €/kWh, which is the price of purchasing electricity from the grid. Thus, the demand-dependent electricity cost resolve at which point the energy system switches to buying or selling electricity from/to the grid and where internal electricity generation determines the cost.

The actual cost savings resulting from Step ③ are calculated in Step ④, since Step ③ employs only the approximate demand-dependent energy cost. In Step ④, the energy systems optimize again for the rescheduled energy demand, leading to final overall cost savings of 8.5 % compared to the sequential optimization. Thus, the largest cost savings (5.3 %) are reached by the rescheduling and coordination of the production systems in Step ③ in combination with the rescheduling of the energy systems in Step ④. Hence, the demand-dependent energy costs enable the main cost savings by rescheduling and coordinating the production systems.

Table 6.1: Cost savings of different optimization approaches in % compared to sequential optimization for the instances of case study I.

instance	1	2	3	4	5	6	7	8	9	10	∅
integrated	9.79	10.63	11.31	10.65	10.63	9.68	10.16	11.02	8.64	8.94	10.15
proposed method	8.96	9.56	9.39	8.14	9.36	7.57	7.48	9.46	5.57	7.21	8.22

The proposed method solves the multi-leader multi-follower Stackelberg game in 1389 s. Therein, step ③ needs 4 iterations which take 986 s. The integrated optimization is solved in 325 s. The short calculation time shows that the proposed method can be implemented in daily scheduling.

In case study I, we generated 10 instances with Latin-hypercube sampling (McKay et al., 2000). The instances are generated with variations of ±20 % around the original energy demands of the production systems. For these 10 instances, the proposed method reduces the total production cost on average by 8.2 % compared to the sequential optimization (Table 6.1). The cost savings even correspond to 81 % of the integrated optimization (10.2 %). Thus, the proposed method largely exploits the potential for cost reductions while only exchanging incomplete information.

Additionally to the constant electricity prices, we also apply our method with time-of-use electricity prices. For this purpose, we used the variation of the electricity prices from the German spot market. We used data starting at 15.7.2020, 0 a.m. (Bundesnetzagentur | SMARD.de, 2020). The results are similar to the case with constant electricity prices. The proposed method results in cost savings of 8 % compared to the sequential optimization. The integrated optimization saves 9.3 % compared to the sequential optimization.

6.2.2 Case study II

6.2.2.1 Description

In this case study, we apply the proposed method to a large industrial site with 4 production systems and 2 energy systems. Production systems 1 and 3 are based on the case study from Kallrath (2002) (Figure 3.9), and production systems 2 and 4 are based on the case study from Kondili et al. (1993) (Figure 3.2). Similar to case study I, we added electricity and heat demands to the tasks.

The product demands in Table 6.2 need to be fulfilled at the end of the time horizon of 30 h. Again, the models of the two energy systems are based on the model by Voll

Table 6.2: Product demand of the production systems in case study II.

production system	1	2	3	4
product demand	State 16: 20 t State 17: 20 t State 19: 20 t	State 7: 300 t State 10: 550 t	State 15: 20 t State 17: 20 t State 18: 20 t	State 7: 500 t State 10: 250 t

et al. (2013). Energy system 1 has 4 boilers (3.5 MW, 3 MW, 2 MW, 1 MW) and 2 combined-heat-and-power engines (3 MW, 2 MW). In the sequential optimization, energy system 1 supplies production systems 3 and 4. Energy system 2 has 6 boilers (5 MW, 4 MW, 1.5 MW, 1.5 MW, 0.5 MW, 0.5 MW) and 4 combined-heat-and-power engines (2.5 MW, 1.5 MW, 1.5 MW, 1 MW). In the sequential optimization, energy system 2 supplies production systems 1 and 2. The prices to buy gas, buy electricity, and sell electricity are the same as in case study I. The optimization problems are formulated in GAMS 24.7.3 and solved with the same solvers as case study I (Section 6.2.1.1). The maximal number of iterations in Step ③ is set to 10.

6.2.2.2 Results

The proposed method saves 2.8 % in cost compared to the sequential optimization (Figure 6.8) and is solved within 66,749 s. The integrated optimization saves 4.1 %. (Figure 6.8). Thus, the proposed method reaches 68 % of the cost savings of an integrated optimization. Again, the cost savings results from large savings in energy costs by 6.1 %, while production cost increase by 1.3 % (Figure 6.8). As in case study I, the total costs of the production systems are decreased as a sum of energy cost and production cost.

The proposed method changes the total energy demand compared to the sequential optimization (Figure 6.9): The peak demand for electricity is decreased by 20.8 % and the peak demand for heat is increased by 12.8 %. Again, the overall electricity demand decreases (11.9 %) while the heat demand increases (2.6 %). Since the heat demand is much higher than the electricity demand, the overall energy demand changes only slightly (−2.9 %). The shift from heat to electricity is possible because the production systems change their production schedules to produce the desired products. Again, the electricity supply by the electricity grid decreases (sequential 9.6 %; proposed method 7.3 %). Thus, the utilization of the on-site energy system increases.

For case study II, after Substep a2 in Step ②, we reach cost savings of 0.7 % compared to the sequential optimization. After Substep a4, the cost savings are nearly similar (0.7 %). Thus, as in case study I, the heuristic allocation already provides a

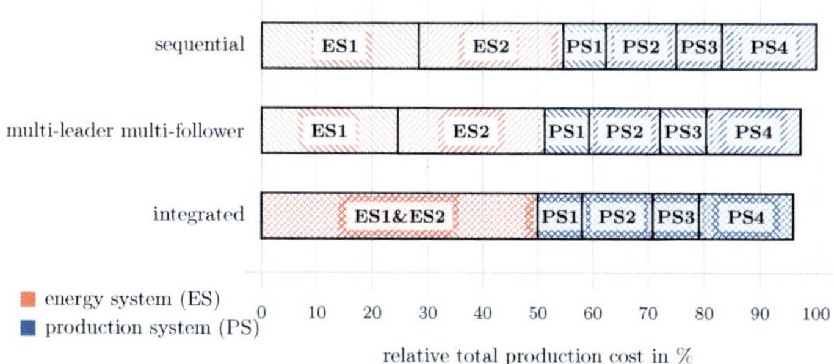

Figure 6.8: Cost in case study II for the different optimization approaches: the common sequential optimization between each production system and the corresponding energy system (sequential = 100 %), the multi-leader multi-follower Stackelberg game solved by the proposed coordination method (multi-leader multi-follower) and the integrated optimization of all systems (integrated). In the integrated optimization, the cost of the energy systems cannot be assigned to the different systems because cost for additional electricity cannot be allocated unambiguously.

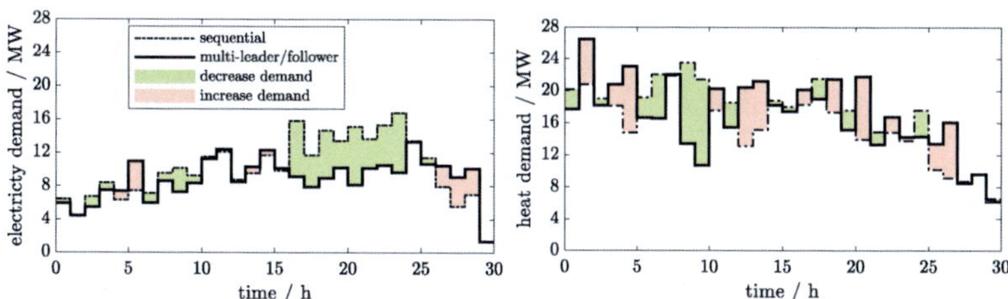

Figure 6.9: Case study II: energy demand of the production systems for the sequential optimization and for the proposed method to coordinate the multi-leader multi-follower Stackelberg game (electricity (left) and heat (right) demand).

good solution. Still, we believe step a4 is crucial for settings with energy systems that are very different in size and employed technologies. After coordination between the production systems using the demand-dependent energy cost (Step ③ and ④), the overall cost savings are 2.8 %. Thus, as in case study I, the rescheduling and coordination of the production systems in Step ③ and the rescheduling of the energy

Table 6.3: Cost savings of different optimization approaches in % compared to sequential optimization for the instances of case study II.

instance	1	2	3	4	5	6	7	8	9	10	∅
integrated	3.52	3.17	5.48	4.02	4.04	3.91	5.01	3.48	4.60	4.13	4.14
proposed method	2.65	2.73	5.31	1.39	3.85	2.68	3.98	0.96	3.38	3.64	3.06

systems enable additional cost savings. Consequently, Steps ① and ② are the basis for expanding the cost savings in Steps ③ and ④.

Again, we generate 10 instances of the energy demands by Latin-hypercube sampling (McKay et al., 2000). The instances are generated with variations of ±20 % around the original energy demands of the production systems. For the instances, the proposed method reduces the cost on average by 3.1 % (Table 6.3) and, thus, is close to the integrated optimization (4.1 %). Consequently, 74 % of the potential cost reduction by integrated optimization is reached.

Also, for case study II, we apply our method for time-of-use electricity prices. We use the German spot market's electricity price variations, starting at 15.7.2020, 0 a.m. (Bundesnetzagentur | SMARD.de, 2020). With these time-of-use electricity prices, the proposed method saves 2.8 % of the overall cost compared to the sequential optimization. The integrated optimization saves 4.3 %.

The case studies show the benefits of the proposed method. However, since the proposed method is a heuristic that cannot guarantee the optimal solution, the proposed method's benefits differ for both case studies. In case study II, the additional production systems supply products not demanded in case study I. Furthermore, the energy systems have more and different components. Still, in both case studies, improved production schedules are identified resulting in energy demands with significantly lower energy cost. The lower energy cost outweighs the slightly increased production cost.

6.3 Conclusions

In this chapter, we extend the method proposed in Chapter 5 with methods for coordination among production systems and coordination among energy systems. With these methods, we can solve the Stackelberg game between multiple production systems and multiple energy systems heuristically. The overall method reduces the total production cost of multiple production systems in a multi-leader multi-follower Stackelberg game.

The proposed method exchanges only incomplete information, i.e., energy demand and demand-dependent energy cost. Between energy and production systems, only one feedback iteration is employed.

The method is tested for two case studies. In the case studies, the total production cost is reduced by 8.5 % and 2.8 % compared to the common sequential optimization. The method realizes 88 % and 68 % of the potential cost reduction by an integrated optimization. The cost savings by the proposed method are obtained by revised production schedules. The revised production schedules lead to slightly increased production cost but significantly decreased energy cost. The proposed method is further validated by large computational studies of 10 instances for both case studies.

The proposed method significantly reduces total production cost while only exchanging incomplete information between energy and production systems in one feedback iteration. Thus, the method is well suited for practical applications.

In Chapter 5 and in Chapter 6, we present methods that solve Stackelberg games using incomplete information. The production systems as leaders can only approximate the behavior of the energy systems. Thus, the identification of the optimal solution for the Stackelberg game is not guaranteed. However, if the production systems know the optimization problem of the energy system, the Stackelberg game can be formulated as a bilevel problem. By solving the bilevel problem, the best solution for the production system under the given situation is identified. Thus, in the following chapter, a method to solve the scheduling of energy and production systems in a bilevel problem is presented.

Chapter 7

Bilevel optimization for joint scheduling of energy and production systems

In this chapter, we again assume that energy and production system have misaligned objectives but now exchange complete information. Even with complete information, the situation can be described as a Stackelberg game. In consequence, we assume that energy and production system are operated by different companies (Figure 2.1). For this purpose, we formulate a bilevel problem and present a solution algorithm. The algorithm is based on the work of Djelassi et al. (2019) and uses the relevant parts for the bilevel problem at hand. Furthermore, we propose a procedure that determines the set of dependent variables adaptively since the dependent variables are assumed to be known in the original algorithm. In the procedure, we take advantage of the properties of the energy system model.

The chapter is organized as follows: In Section 7.1, we present the bilevel problem. In Section 7.2, the algorithm to solve the bilevel problem is presented as well as the optimization problems used in the algorithm. In Section 7.3, we apply the algorithm to two case studies. With the case studies, we present the performance of the algorithm. In these case studies, we compare the methods for sequential optimization, integrated optimization, coordination with incomplete information exchange, and the bilevel optimization. In Section 7.4, conclusions are presented.

Leenders, L., Hagedorn, D. F., Djelassi, H., Bardow, A., and Mitsos, A. (2022). Bilevel optimization for joint scheduling of production and energy systems. *Optimization & Engineering*.

The author of this thesis contributed to the development and implementation of the method as well as the calculation and interpretation of the results, wrote the draft of the paper and is its principal author.

7.1 Bilevel problem for scheduling energy and production systems

In this section, first, we introduce the bilevel problem for the scheduling of production and energy systems. Second, we analyze the properties of the lower-level problem. With the properties, we devise a procedure that determines an appropriate set of dependent variables adaptively.

7.1.1 Formulation of the bilevel problem

The bilevel problem considers scheduling the production system in the upper level and scheduling the energy system in the lower level. The decision variables of the upper level, i.e., production system, are the continuous variables[1] $\mathbf{x} \in \mathbb{R}^{n_x}$ and the binary variables $\mathbf{y} \in \{0,1\}^{n_y}$. $\mathbf{x}$ includes the energy demand $E_{t,e}^{\text{demand}}$ in each time step t and for each energy form e. We denote the joint upper-level variables as $\mathbf{w} := (\mathbf{x}, \mathbf{y})$. The decision variables of the lower level, i.e., energy system, are the energy supplied by the energy conversion units $\mathbf{V}' \in \mathbb{R}^{n_V}$ and the operational state of the energy conversion units $\mathbf{o}' \in \{0,1\}^{n_o}$. As it is the common notation in bilevel modeling, the decision variables of the lower level are written with a prime symbol $'$. For the representation of the lower-level variables in the upper-level problem, the prime symbol $'$ is omitted. The MILBP has the following form:

$$\min_{\mathbf{w},\mathbf{V},\mathbf{o},E_{t,e}^{\text{demand}}} \quad f^{\text{u}}(\mathbf{w},\mathbf{V},\mathbf{o}) = \mathbf{c}^{\text{PS}^T}\mathbf{w} + \mathbf{c}^{\text{U,ES,V}^T}\mathbf{V} + \mathbf{c}^{\text{U,ES,o}^T}\mathbf{o} \tag{7.1}$$

$$\text{s.t.} \quad \mathbf{A}^{\text{PS}}\mathbf{w} \leq \mathbf{b}^{\text{PS}} \tag{7.2}$$

$$(\mathbf{V},\mathbf{o}) \in \arg\min_{\mathbf{V}',\mathbf{o}'} \quad f^{\text{l}}(\mathbf{V}',\mathbf{o}') = \mathbf{c}^{\text{L,ES,V}^T}\mathbf{V}' + \mathbf{c}^{\text{L,ES,o}^T}\mathbf{o}' \tag{7.3}$$

$$\text{s.t.} \quad \sum_{u\in U} V'_{t,u,e} = E_{t,e}^{\text{demand}}, \forall t \in T, e \in E \tag{7.4}$$

$$\mathrm{V}^{-}_{t,u,e}(o'_{t,u}) \leq V'_{t,u,e} \leq \mathrm{V}^{+}_{t,u,e}(o'_{t,u}), \forall t \in T, u \in U, e \in E \tag{7.5}$$

$$\mathrm{a}_u \cdot o'_{t,u} = \mathrm{b}_u \cdot V'_{t,u,e_1} - V'_{t,u,e_2}, \forall t \in T, (u,e_1,e_2) \in C. \tag{7.6}$$

[1]In this chapter, we write vectors in bold.

The upper-level optimization problem (scheduling of the production system) is presented in Equation (7.1) and Equation (7.2) and the lower-level optimization problem (scheduling of the energy system) in Equations (7.3) - (7.6).

The upper-level objective $f^{\mathrm{u}}(\mathbf{w}, \mathbf{V}, \mathbf{o})$ considers production cost $\mathbf{c}^{\mathrm{PS}^T}\mathbf{w}$ and energy cost $\mathbf{c}^{\mathrm{U,ES,V}^T}\mathbf{V} + \mathbf{c}^{\mathrm{U,ES,o}^T}\mathbf{o}$ to be paid by the production system. The vectors $\mathbf{c}^{\mathrm{PS}^T}$, $\mathbf{c}^{\mathrm{U,ES,V}^T}$ and $\mathbf{c}^{\mathrm{U,ES,o}^T}$ describe the cost resulting from the upper- and lower-level variables. $\mathbf{w}$ are the upper-level variables, and thus, the variables of the production system. $\mathbf{V}$ and $\mathbf{o}$ represent the lower-level variables in the upper level. Again, we model the production system as a MILP and summarize the constraints in matrix form with $\mathbf{A}^{\mathrm{PS}}$ and $\mathbf{b}^{\mathrm{PS}}$ as parameters of the production system. In the case study, we model the production system as a batch production system and use the State-Task-Network formulation (Kondili et al., 1993). However, also other models of the production system could be used in the upper level.

The lower-level objective $f^{\mathrm{l}}(\mathbf{V}', \mathbf{o}')$ considers cost of the energy system $\mathbf{c}^{\mathrm{L,ES,V}^T}\mathbf{V}'$ $+\mathbf{c}^{\mathrm{L,ES,o}^T}\mathbf{o}'$ to supply the production system with energy. $\mathbf{c}^{\mathrm{L,ES,V}^T}$ and $\mathbf{c}^{\mathrm{L,ES,o}^T}$ are vectors for the cost resulting from the lower-level variables. The lower-level constraints consider energy balances, part-load performance, the operational limits of energy conversion units, etc. $\mathbf{V}'$ is a vector of the energy $V'_{t,u,e}$ of energy form e provided by energy conversion unit u in time step t. $\mathbf{o}'$ is a vector of the operational state $o'_{t,u}$ of energy conversion unit u in time step t. $o'_{t,u}$ is a binary variable and equals 1 if energy conversion unit u is operated in time step t. $\mathbf{V}'$ and $\mathbf{o}'$ are the decision variables of the energy system.

The coupling equality constraints between lower and upper level are given by energy balances, stated separately in Equation (7.4). In the coupling equality constraints, upper-level variables are the energy demands of the production system $E^{\mathrm{demand}}_{t,e}$ for each energy form e in time step t. To clarify that the energy demand is an upper-level decision variable, we also write $E^{\mathrm{demand}}_{t,e}$ under the min operator, even if it is an entry of the vector $\mathbf{w}$. Lower-level variables in the coupling equality constraints are the energy $V'_{t,u,e}$ of energy form e provided by energy conversion unit u in time step t.

Again, we linearize the nonlinear unit behavior by piecewise-affine linear functions (Voll et al., 2013). For an easier separation into dependent and independent variables in the proposed algorithm (c.f. Section 7.2.2), we model each segment of the piecewise-affine linear function as a separate energy conversion unit. Their operational states are connected via constraints. In the case study, we use again the energy system model from Voll et al. (2013). However, also other energy system models could be used in the lower-level problem. Still, the part-load performance needs to be modeled by piecewise-affine functions, and all equations need to be linear as we assume this in our proof (Section 7.1.2 and Appendix E).

Here, all inequality constraints of the lower level are reformulated as parametric upper and lower bounds of the entries of $\mathbf{V}'$. Thereby, $\mathrm{V}^{-}_{t,u,e}(o'_{t,u})$ and $\mathrm{V}^{+}_{t,u,e}(o'_{t,u})$ are affine linear functions (Equation (7.5)). Equation (7.6) models the connection between two energy forms. These connections appear only for specific energy conversion units, e.g., combined-heat-and-power engines which supply electricity and heat simultaneously. These energy conversion units u with the corresponding two energy forms e_1 and e_2 are summarized in the set C. a_u and b_u are parameters to describe the connection between the energy forms.

7.1.2 Properties of the lower-level problem

In the energy system model, we model no constraints to connect time steps. As a result, the lower-level decisions at one time step are independent of the decisions at all other time steps. Therefore and by linearity of the lower-level objective, the lower-level problem can be decomposed in $|T|$ lower-level problems, one for each time step. The optimization problem of the lower level for a single time step t is:

$$\min_{\mathbf{V}_t,\mathbf{o}_t} f^{\mathrm{l}}_t(\mathbf{V}_t,\mathbf{o}_t) = \mathbf{c}_t^{\mathrm{L,ES,V}^T}\mathbf{V}_t + \mathbf{c}_t^{\mathrm{L,ES,o}^T}\mathbf{o}_t \tag{7.7}$$

$$\text{s.t.} \sum_{u\in U} V_{t,u,e} = E^{\text{demand}}_{t,e}, \forall e \in E \tag{7.8}$$

$$\mathrm{V}^{-}_{t,u,e}(o_{t,u}) \le V_{t,u,e} \le \mathrm{V}^{+}_{t,u,e}(o_{t,u}), \forall u \in U, e \in E \tag{7.9}$$

$$\mathrm{a}_u \cdot o_{t,u} = \mathrm{b}_u \cdot V_{t,u,e_1} - V_{t,u,e_2}, \forall (u,e_1,e_2) \in C. \tag{7.10}$$

The decision variables of the lower level are the output energy of the energy conversion units $\mathbf{V}_t$ and the operational state of the energy conversion units $\mathbf{o}_t$.

In Section 7.2, we present the algorithm to solve the bilevel problem from Equation (7.1)-(7.6). In the algorithm, we use the property that an optimal solution always exists for the lower-level problem where all except one energy conversion unit per energy form e are operated at one of their operational limits. We prove this property in the following proposition and the proposition in Appendix E.

Proposition 1. *Let the lower-level problem be given for a single energy form e and one time step t as*

$$\begin{aligned}\min_{\boldsymbol{V}_{t,e},\boldsymbol{o}_t} f^l_{t,e}(\boldsymbol{V}_{t,e},\boldsymbol{o}_t) &= \boldsymbol{c}^{L,ES,V^T}_{t,e}\boldsymbol{V}_{t,e} + \boldsymbol{c}^{L,ES,o^T}_{t}\boldsymbol{o}_t\\ s.t. \sum_{u\in U} V_{t,u,e} &= E^{demand}_{t,e}\\ V^-_{t,u,e}(o_{t,u}) \leq V_{t,u,e} &\leq V^+_{t,u,e}(o_{t,u}), \forall u \in U.\end{aligned} \tag{7.11}$$

Therein, $\boldsymbol{V}_{t,e}$ is the vector of energy supplied by the energy conversion units, and $\boldsymbol{o}_t$ is the vector of the operational state of each energy conversion unit. Note that the constraints to connect the energy forms (Equation (7.10)) are omitted because only one energy form is considered.

For this problem, there exists an optimal solution in which at most one energy conversion unit is not operated at one of its operating limits. More precisely, let Problem (7.11) be feasible and let $(\boldsymbol{o}^*_t, \boldsymbol{V}^*_{t,e})$ be globally optimal in Problem (7.11). Then, there exists a solution $(\boldsymbol{o}^*_t, \tilde{\boldsymbol{V}}_{t,e})$ that is also globally optimal in Problem (7.11) and for which there exists at most one $u \in U$ such that

$$V^-_{t,u,e}(\boldsymbol{o}^*_t) < \tilde{V}_{t,u,e} < V^+_{t,u,e}(\boldsymbol{o}^*_t). \tag{7.12}$$

Proof. Consider the case that $(\mathbf{o}^*_t, \mathbf{V}^*_{t,e})$ is given such that for zero or one $u \in U$ the following equation holds:

$$\mathrm{V}^-_{t,u,e}(\mathbf{o}^*_t) < V^*_{t,u,e} < \mathrm{V}^+_{t,u,e}(\mathbf{o}^*_t). \tag{7.13}$$

Then, the result is proven immediately with $\tilde{\mathbf{V}}_{t,e} = \mathbf{V}^*_{t,e}$.

Now consider instead that $u_1, u_2 \in U, u_1 \neq u_2$ be given such that

$$\mathrm{V}^-_{t,u,e}(\mathbf{o}^*_t) < V^*_{t,u,e} < \mathrm{V}^+_{t,u,e}(\mathbf{o}^*_t), \quad \forall u \in \{u_1, u_2\} \tag{7.14}$$

and

$$V^*_{t,u,e} = \mathrm{V}^-_{t,u,e}(\mathbf{o}^*_t) \vee V^*_{t,u,e} = \mathrm{V}^+_{t,u,e}(\mathbf{o}^*_t), \quad \forall u \in U \setminus \{u_1, u_2\}. \tag{7.15}$$

Then, it follows from construction of Equation (7.11) that all elements of the set

$$\begin{aligned}M = \{(\mathbf{o}^*_t, \mathbf{V}_{t,e}) | V_{t,u,e} &= V^*_{t,u,e}, \forall u \in U \setminus \{u_1, u_2\}\\ &\wedge \mathrm{V}^-_{t,u,e}(\mathbf{o}^*_t) \leq V_{t,u,e} \leq \mathrm{V}^+_{t,u,e}(\mathbf{o}^*_t), \forall u \in \{u_1, u_2\}\\ &\wedge V_{t,u=u_1,e} + V_{t,u=u_2,e} = V^*_{t,u=u_1,e} + V^*_{t,u=u_2,e}\}\end{aligned} \tag{7.16}$$

are feasible in Equation (7.11). $\mathbf{V}^*_{t,e}$ is an optimal solution of a linear program on a facet of the feasible set. Accordingly, all points on that facet (points in M) are optimal in Equation (7.11).

The point $(\mathbf{o}^*_t, \tilde{\mathbf{V}}_{t,e})$ with $\tilde{V}_{t,u,e} = V^*_{t,u,e}, \forall u \in U \setminus \{u_1, u_2\}$ and

$$\begin{aligned} \tilde{V}_{t,u=u_1,e} &= V^*_{t,u=u_1,e} + \min\{V^+_{t,u=u_1,e}(\mathbf{o}^*_t) - V^*_{t,u=u_1,e}, V^*_{t,u=u_2,e} - V^-_{t,u=u_2,e}(\mathbf{o}^*_t)\} \\ \tilde{V}_{t,u=u_2,e} &= V^*_{t,u=u_2,e} - \min\{V^+_{t,u=u_1,e}(\mathbf{o}^*_t) - V^*_{t,u=u_1,e}, V^*_{t,u=u_2,e} - V^-_{t,u=u_2,e}(\mathbf{o}^*_t)\} \end{aligned} \tag{7.17}$$

lies within M and satisfies either $\tilde{V}_{t,u=u_1,e} = V^+_{t,u=u_1,e}(\mathbf{o}^*_t)$ or $\tilde{V}_{t,u=u_2,e} = V^-_{t,u=u_2,e}(\mathbf{o}^*_t)$, proving the desired property. Note that in an optimal solution where two energy conversion units are not operated at their operational limits, the cost factors $c^{\mathrm{L,ES,V}}_{t,u,e}$ for these energy conversion units are equal. Otherwise, a change in the supplied energy by the energy conversion units could result in a lower objective function value.

Finally, if there are more than two $u \in U$ with $\mathrm{V}^-_{t,u,e}(\mathbf{o}^*_t) < V^*_{t,u,e} < \mathrm{V}^+_{t,u,e}(\mathbf{o}^*_t)$, the above construction can be applied successively until the same result is reached. □

We proved that there exists a solution for which only one energy conversion unit is not operated at one of its operational limits if there is only one form of energy e. More generally, in Appendix E, we show that the number of energy conversion units not operated at one of their operational limits can be reduced to the number of energy forms e or below.

7.2 Algorithm to solve the bilevel problem

In Section 7.2.1-7.2.3, we first present the algorithm, and second, in Section 7.2.4, we present the optimization problems considered in the algorithm: lower-bounding problem, lower-level problem, and auxiliary problem.

7.2.1 Algorithm

The proposed algorithm is based on the algorithm from Djelassi et al. (2019). The algorithm from Djelassi et al. (2019) is based on Mitsos et al. (2008) and Mitsos (2010). Djelassi et al. (2019) use discretization points from optimal lower-level solutions for representing the lower-level objective in the upper-level problem. We assume that the lower-level variables are split into sets of dependent and independent variables. These sets are then treated differently in the discretization: Independent variables are fixed, and dependent variables are determined by the equality constraints.

In the bilevel problem at hand, the choices for dependent and independent variables change for each discretization point. In the previous section and in the appendix, we prove that there are at most as many energy conversion units not operated at their operational limits as there are energy forms e considered. Thus, we can always set the energy provided by the energy conversion units not operated at their operational limits as the dependent variables. The energy supply by the energy conversion units operated at their operational limits are then the independent variables. Hence, there are at most as many dependent variables in an optimal solution point of the lower-level problem as energy forms are supplied by the energy system (Section 7.1.2).

We use these properties to identify discretization points from lower-level solutions, which means we identify independent and dependent variables. Thereby, the algorithm can solve the MILBP of energy and production system scheduling stated in Section 7.1.1.

In the following Section 7.2.2, we introduce the discretization points used in the algorithm. Afterward, we present the steps of the algorithm in Section 7.2.3.

7.2.2 Discretization points

A discretization point defines the independent and dependent variables for an optimal lower-level solution. Therein, the independent variables are fixed, and the dependent variables are determined by the equality constraints. Discretization points are used to represent optimal lower-level solutions in the upper-level problem.

In the bilevel problem of scheduling energy and production systems, the energy balances are the only coupling constraints between the upper- and the lower-level problem. The energy balance is given in Equation (7.4) and repeated here:

$$\sum_{u \in U} V_{t,u,e} = E_{t,e}^{\text{demand}}, \forall t \in T, e \in E. \tag{7.18}$$

The energy demands $E_{t,e}^{\text{demand}}$ are the upper-level variables in the energy balance, and the energy $V_{t,u,e}$ provided by each energy conversion unit u are the lower-level variables. Thus, the only variable from the upper-level problem in the lower-level problem are the energy demands $E_{t,e}^{\text{demand}}$. Since discretization points are used to define optimal lower-level solutions in the upper level, discretization points are defined for given energy demands $E_{t,e}^{\text{demand}}$. Thus, discretization points indicate the energy supplied by each energy conversion unit if a particular energy demand occurs.

In Section 7.1.2 and Appendix E, we proved that there exists an optimal solution in which the number of energy conversion units not operated at one of their operational limits is less or equal to the number of energy forms e considered. We identify this optimal solution in the auxiliary problem (Section 7.2.4.3). This optimal solution is only valid for a specific energy demand. However, suppose we vary this energy demand, keep the power of the energy conversion units operated at their operational limits fixed, and compensate for the variation by adjusting the power of the energy conversion units not operated at their operational limits. In that case, the solution is still feasible until an energy conversion unit exceeds its operational limits. In this range of validity, the lower-level objective function value identified by the discretization point provides a lower-level feasible solution and, therefore, an upper bound of the lower-level objective function. Still, within the range of validity of a discretization point, another discretization point can give a tighter upper bound.

We call the energy conversion units not operated at their operational limits *free energy conversion units*. These units define the dependent variables. The energy conversion units operated at their operational limits are named *discretized energy conversion units* and are the independent variables. For discretized energy conversion units, the operation is fixed in the discretization point k. These energy conversion units are either idle (index-set: D_k^{zero}) or operated at their minimal or maximal load (index-sets: D_k^{min} and D_k^{max}). Thus, the energy $V_{k,t,u,e}^{\text{D}}$ supplied by these discretized energy conversion units is fixed in the discretization point k:

$$V_{k,t,u=d,e}^{\text{D}} = 0, \qquad \forall k \in K, t \in T, d \in D_k^{\text{zero}}, e \in E \tag{7.19}$$

$$V_{k,t,u=d,e}^{\text{D}} = \text{V}_{u=d,e}^{\text{min}}, \qquad \forall k \in K, t \in T, d \in D_k^{\text{min}}, e \in E \tag{7.20}$$

$$V_{k,t,u=d,e}^{\text{D}} = \text{V}_{u=d,e}^{\text{max}}, \qquad \forall k \in K, t \in T, d \in D_k^{\text{max}}, e \in E. \tag{7.21}$$

Here, $\text{V}_{u,e}^{\text{min}}$ and $\text{V}_{u,e}^{\text{max}}$ are the minimal and maximal load of energy conversion unit u, respectively.

For each discretization point k, time point t, and energy form e, the supplied energy of one energy conversion unit u is not fixed, and thus, is free. These free energy conversion units are summarized by the set $d_{k,f}^{\text{free}}$. The index f counts the free energy conversion units for each discretization point k.

To fulfill each energy demand, the supplied energy of the free energy conversion units $V_{k,t,u=d_{k,f}^{\text{free}},e}^{\text{D}}$ is not limited. This unlimited supplied energy can lead to supplied energy values $V^{\mathrm{D}}_{k,t,u=d^{\mathrm{free}}_{k,f},e}$ outside the operational limits of the free energy conversion unit. Outside the operational limits, the upper bound on the lower-level objective function is not considered anymore (c.f. Appendix D, Equation (7.22) and Equation (7.23)).

7.2.3 Steps of the algorithm

Algorithm 1 solves the MILBP by a lower-bounding problem, lower-level problem, and auxiliary problem. The lower-bounding problem is a finite (discretization-based) relaxation of the MILBP. The lower-level problem is solved for fixed upper-level variables. The lower-level optimal solution is used to evaluate the upper-level objective to generate an upper bound. This upper bound is valid in our case since there are no upper-level constraints depending on the lower-level variables. Bilevel optimality of the upper-bounding solution is assessed based on the lower bound provided by the lower-bounding problem. In an auxiliary problem, discretization points are identified from the solutions of the lower-level problem. The algorithm repeatedly solves these problems while the lower-bounding problem is tightened by adding discretization points until the optimality tolerance ϵ^{u} between upper and lower bound is reached.

Our algorithm exploits the structure of the lower-level problem. In the lower-level problem, all constraints are written for each time step and no constraint connects the time steps. The objective function only sums the cost of each time step. Thus, the lower-level problem can be decomposed into individual optimization problems for each time step t. Each of these individual optimization problems is parameterized only by its energy demand $E^{\mathrm{demand}}_{t,e}$. The solution obtained at one time step constitutes a valid discretization point for all time steps because discretization points are defined for energy demands in a single time step which can occur in multiple time steps. Thus, probably fewer iterations are needed to solve the bilevel problem. In general, the algorithm might also be suitable to consider constraints that connect time steps in the lower-level constraints. In this case, each discretization point defines the lower-level solution for a specific energy-demand curve of the whole time horizon.

The algorithm terminates in finite time because there is a finite number of possible discretization points. The number is finite because there is a maximum number of combinations of free energy conversion units and discretized energy conversion units. In the worst case, an enumeration of all the discretization points would be necessary. However, in our applications of the algorithm, we observe that the algorithm does much better in practice and only needs a few iterations.

Algorithm 1: Algorithm to solve bilevel problem of scheduling energy and production system.

1 Set $LBD \leftarrow -\infty$, $UBD \leftarrow \infty$, $\epsilon^{\mathrm{u}} > 0$;
2 Solve lower-bounding problem to obtain $f^{\mathrm{u,LBD}}$ and $\mathbf{w}^{\mathrm{LBD}}$;
3 Set $LBD \leftarrow f^{\mathrm{u,LBD}}$;
4 Solve lower-level problem for $\mathbf{w}^{\mathrm{LBD}}$ to obtain $f^{\mathrm{l,LLP}}(\mathbf{w}^{\mathrm{LBD}}, \mathbf{V}^{\mathrm{LLP}}, \mathbf{o}^{\mathrm{LLP}})$, $\mathbf{V}^{\mathrm{LLP}}$ and $\mathbf{o}^{\mathrm{LLP}}$;
5 Evaluate upper-level objective for upper-level solution $\mathbf{w}^{\mathrm{LBD}}$ and lower-level solution $\mathbf{V}^{\mathrm{LLP}}$ and $\mathbf{o}^{\mathrm{LLP}}$ to obtain $f^{\mathrm{u}}(\mathbf{w}^{\mathrm{LBD}}, \mathbf{V}^{\mathrm{LLP}}, \mathbf{o}^{\mathrm{LLP}})$;
6 **if** $f^{u}(\boldsymbol{w}^{LBD}, \boldsymbol{V}^{LLP}, \boldsymbol{o}^{LLP}) < UBD$ **then**
7 | Set $(\mathbf{w}^{*}, \mathbf{V}^{*}, \mathbf{o}^{*}) \leftarrow (\mathbf{w}^{\mathrm{LBD}}, \mathbf{V}^{\mathrm{LLP}}, \mathbf{o}^{\mathrm{LLP}})$, $UBD \leftarrow f^{\mathrm{u}}(\mathbf{w}^{\mathrm{LBD}}, \mathbf{V}^{\mathrm{LLP}}, \mathbf{o}^{\mathrm{LLP}})$;
8 **end**
9 **if** $UBD - LBD \leq \epsilon^{u}$ **then**
10 | terminate with solution $(\mathbf{w}^{*}, \mathbf{V}^{*}, \mathbf{o}^{*})$;
11 **end**
12 Solve auxiliary problem for $\mathbf{w}^{\mathrm{LBD}}$ and $f^{\mathrm{l,LLP}}(\mathbf{w}^{\mathrm{LBD}}, \mathbf{V}^{\mathrm{LLP}}, \mathbf{o}^{\mathrm{LLP}})$ to obtain $\mathbf{V}^{\mathrm{aux}}, \mathbf{o}^{\mathrm{aux}}$;
13 Extend set of discretization points K in the lower-bounding problem by new discretization points identified from $\mathbf{V}^{\mathrm{aux}}$ and $\mathbf{o}^{\mathrm{aux}}$;
14 Go to Line 2;

7.2.4 Subproblems of the algorithm

In this section, we present the optimization problems considered in the algorithm: lower-bounding problem, lower-level problem, and auxiliary problem. To explain the underlying ideas more specifically, we present the algorithm for the supply of two energy forms e, i.e., heat and electricity. These energy forms are assumed to be supplied by boilers, combined-heat-and-power engines, and electricity grids.

7.2.4.1 Lower-bounding problem

The lower-bounding problem is the upper-level optimization problem (production system) while considering the constraints of the lower-level problem (energy system) and the lower-level objective by discretization points.

In the 1st iteration, and thus, without discretization points, the lower-bounding problem is composed from Equations (7.1), (7.2), (7.4), (7.5), and (7.6). Thus, in the 1st iteration, the lower-bounding problem corresponds to the integrated scheduling, where the production system has full control over the energy system.

After the 1st iteration, the optimal solutions of lower-level problems are considered by discretization points K. The lower-level objective resulting from a discretization point $\mathrm{f}_{k,t}^{\mathrm{l,Discr.\ Point}}$ is an upper bound on the lower-level objective $f_t^{\mathrm{l,LBD}}$ in the discretization points range of validity (Section 7.2.2). We model this upper bound by a big-M formulation:

$$f_t^{\mathrm{l,LBD}} \leq \mathrm{f}_{k,t}^{\mathrm{l,Discr.\ Point}} + \mathrm{M}^{\mathrm{obj}} \cdot (1 - \beta_{k,t}^{\mathrm{obj}}), \forall k \in K, t \in T. \tag{7.22}$$

The objective function value is only constrained if the binary variable $\beta_{k,t}^{\mathrm{obj}}$ equals 1. $\beta_{k,t}^{\mathrm{obj}}$ equals 1 if the supplied energy of the free energy-conversion units is within the range of validity of the discretization point. Note that for a specific energy demand, multiple discretization points can give an upper bound on the objective function.

The binary variables $\beta_{k,t,f}^{\mathrm{lower}}$ and $\beta_{k,t,f}^{\mathrm{upper}}$ identify if the supplied energy of the free energy conversion unit $V_{k,t,u=d_{k,f}^{\mathrm{free}},e}^{\mathrm{D}}$ is lower or higher than the operational limits. Thus, Equation (7.23) ensures that the upper bound on the objective function in Equation (7.22) is used if the supplied energy of the free energy conversion units is within their operational limits:

$$\sum_{f \in \{1,2\}} (\beta_{k,t,f}^{\mathrm{upper}} + \beta_{k,t,f}^{\mathrm{lower}}) + \beta_{k,t}^{\mathrm{obj}} \geq 1, \forall k \in K, t \in T. \tag{7.23}$$

In Appendix D, equations are stated to identify if the supplied energy of the free energy conversion units is within the operational limits.

7.2.4.2 Lower-level problem (LLP)

The lower-level problem is the operation optimization of the energy system for a given energy demand. The lower-level problem can be solved independently for each time step t because there are no constraints that connect time steps in the lower-level. The lower-level problem is already stated in Equation (7.7)-(7.10). Compared to the algorithm in Djelassi et al. (2019), we do not have to evaluate in an upper-bounding problem if the lower-level solution is feasible for the upper level. The evaluation is not necessary since there are no lower-level variables in the constraints of the upper level. Hence, we can directly use the solutions of the lower-level problem to evaluate the upper-level objective.

7.2.4.3 Auxiliary problem

The auxiliary problem identifies the discretization points from each lower-level solution. In the discretization points, as few as possible energy conversion units are not operated at their operational limits. Thus, the objective of the auxiliary problem is to minimize the number of free energy conversion units while retaining the objective of the lower-level solution. The auxiliary problem considers the constraints of the lower-level problem. Because no constraints connecting time steps are considered in the lower-level problem, an auxiliary problem can be solved for each time step.

For each energy conversion unit, we define the 4 binary variables $\theta_{t,u}^{\text{min}}$, $\theta_{t,u}^{\text{max}}$, $\theta_{t,u}^{\text{zero}}$, and $\theta_{t,u}^{\text{free}}$. $\theta_{t,u}^{\text{min}}$ and $\theta_{t,u}^{\text{max}}$ identify if the energy conversion unit is operated at its minimal or maximal operational limit, respectively. $\theta_{t,u}^{\text{zero}}$ identifies if the energy conversion unit is idle. $\theta_{t,u}^{\text{free}}$ identifies if the energy conversion unit is operated between its operational limits.

The objective function of the auxiliary problem

$$\min_{\mathbf{V}_t^{\text{aux}},\mathbf{o}_t^{\text{aux}}} \sum_{u \in U} \theta_{t,u}^{\text{free}} \tag{7.24}$$

is the minimization of the number of free energy conversion units in time step t.

In the constraints, the objective function value of the lower-level problem calculated with the variables of the auxiliary problem $f_t^{\text{l,aux}}$ needs to be equal or lower than the objective function value from the lower-level problem $\text{f}_t^{\text{l,LLP}}$:

$$f_t^{\text{l,aux}} \leq \text{f}_t^{\text{l,LLP}}. \tag{7.25}$$

The constraints of the lower-level problem need to hold in the auxiliary problem (Equation (7.8) - (7.10)). Additional constraints define the 4 binary variables $\theta_{t,u}^{\text{min}}$, $\theta_{t,u}^{\text{max}}$, $\theta_{t,u}^{\text{zero}}$, and $\theta_{t,u}^{\text{free}}$.

An energy conversion unit u is either operated at its operational limits, operated between its operational limits (free), or idle. Thus, for the sum of the binary variables we can state:

$$\theta_{t,u}^{\text{min}} + \theta_{t,u}^{\text{max}} + \theta_{t,u}^{\text{zero}} + \theta_{t,u}^{\text{free}} = 1, \forall u \in U. \tag{7.26}$$

As stated at the beginning of Section 7.2.4, the equations to identify the energy conversion units at their operational limits are exemplary shown for electricity grids, boilers, and combined-heat-and-power engines. Electricity grids are an example of energy conversion units without an operational limit. Boilers and combined-heat-and-power engines are examples of energy conversion units with upper and lower operational limits. Combined-heat-and-power engines are also an example of energy conversion units connecting two energy forms.

We model two electricity grids, one for the supply of electricity and one for the feed-in of electricity into the grid. For the electricity grids, the 4 binary variables are defined in the following: The electricity grids have no upper operational limits, and zero is the lower operational limit of the electricity grids. Thus, $\theta_{t,u}^{\max}$ and $\theta_{t,u}^{\text{zero}}$ are defined to be zero:

$$\theta_{t,u}^{\max} = 0, \forall u \in D^{\text{Grid}} \tag{7.27}$$

$$\theta_{t,u}^{\text{zero}} = 0, \forall u \in D^{\text{Grid}}. \tag{7.28}$$

D^{Grid} is a set containing all electricity grids.

We identify if the electricity grid is at the lower operational limit ($\theta_{t,u}^{\min} = 1$) by:

$$V_{t,u,e=\text{el}}^{\text{aux}} \leq (1 - \theta_{t,u}^{\min}) \cdot \mathrm{M}_{e=\text{el}}^{\max}, \forall u \in D^{\text{Grid}}. \tag{7.29}$$

$\mathrm{M}_{e=\text{el}}^{\max}$ is a sufficiently large number such that the electricity demand from the electricity grid or the electricity supply to the electricity grid never exceeds this number. Thus, an electricity grid is identified as the free energy conversion unit for electricity ($\theta_{t,u}^{\text{free}} = 1$) if the energy $V_{t,u,e=\text{el}}^{\text{aux}}$ is greater than 0.

The constraints to define the 4 binary variables for boilers and combined-heat-and-power engines are given in the following: Heat and electricity outputs are connected for a combined-heat-and-power engine. Thus, if a combined-heat-and-power engine is operated at its operational limit in heat supply, also the operational limit in the electricity supply is reached. Consequently, we identify if a combined-heat-and-power engine is operated at its operational limit only for one energy form, i.e., heat. We identify if the supplied energy of a combined-heat-and-power engine or a boiler is at its lower operational limit by:

$$\begin{aligned} V_{t,u,e=\text{heat}}^{\text{aux}} \leq \mathrm{V}_{t,u,e=\text{heat}}^{\max} - \theta_{t,u}^{\min} \cdot (\mathrm{V}_{t,u,e=\text{heat}}^{\max} - \mathrm{V}_{t,u,e=\text{heat}}^{\min}), \\ \forall u \in D^{\text{CHP}} \cup D^{\text{Boiler}}. \end{aligned} \tag{7.30}$$

If $\theta_{t,u}^{\min}$ equals 0, the supplied energy $V_{t,u,e=\text{heat}}^{\text{aux}}$ needs to be lower or equal than the heat supply in maximal load $\mathrm{V}_{t,u,e=\text{heat}}^{\max}$. If $\theta_{t,u}^{\min}$ equals 1, the supplied energy $V_{t,u,e=\text{heat}}^{\text{aux}}$ needs to be lower or equal than the heat supply in minimal part-load $\mathrm{V}_{t,u,e=\text{heat}}^{\min}$. Since the other constraints require that if a unit is operated, the supplied energy is greater or equal than the minimal part-load, the unit is operated at its minimal operational limit. D^{CHP} and D^{Boiler} are sets with all combined-heat-and-power engines and boilers, respectively.

If a combined-heat-and-power engine or a boiler is operated at its upper operational limit $\text{V}^{\max}_{t,u,e=\text{heat}}$ is identified by:

$$V^{\text{aux}}_{t,u,e=\text{heat}} \geq \text{V}^{\max}_{t,u,e=\text{heat}} \cdot \theta^{\max}_{t,u}, \quad \forall u \in D^{\text{CHP}} \cup D^{\text{Boiler}}. \tag{7.31}$$

A similar explanation as for Equation (7.30) applies for the upper operational limit.

Whether a combined-heat-and-power engine or a boiler is idle is identified by checking if the supplied energy is within the upper and lower operational limit:

$$V^{\text{aux}}_{t,u,e=\text{heat}} \geq (1 - \theta^{\text{zero}}_{t,u}) \cdot \text{V}^{\min}_{t,u,e=\text{heat}}, \quad \forall u \in D^{\text{CHP}} \cup D^{\text{Boiler}} \tag{7.32}$$

$$V^{\text{aux}}_{t,u,e=\text{heat}} \leq (1 - \theta^{\text{zero}}_{t,u}) \cdot \text{V}^{\max}_{t,u,e=\text{heat}}, \quad \forall u \in D^{\text{CHP}} \cup D^{\text{Boiler}}. \tag{7.33}$$

Thus, if the heat supply $V^{\text{aux}}_{t,u,e=\text{heat}}$ is between the operational limits, the binary variables $\theta^{\min}_{t,u}$, $\theta^{\max}_{t,u}$, and $\theta^{\text{zero}}_{t,u}$ are zero and Equation (7.26) sets $\theta^{\text{free}}_{t,u}$ to 1. In this case, the corresponding unit is identified as a free energy conversion unit for heat.

7.2.4.4 Integration of auxiliary problem in lower-level problem

It should be mentioned that the auxiliary problem can also be integrated into the lower-level problem if the number of free energy conversion units is known. For this purpose, the following constraints should be added:

$$\sum_{u \in U} \theta^{\text{free}}_{t,u} \leq \text{M}^{\text{free}}, \forall t \in T. \tag{7.34}$$

Again, $\theta^{\text{free}}_{t,u}$ identifies if an energy conversion unit u is not operated at its operational limits ($\theta^{\text{free}}_{t,u}$=1). The sum over $\theta^{\text{free}}_{t,u}$ is constrained by the number of allowed free energy conversion units M^{free}. The energy system in the case studies can supply 2 energy forms, and the maximum number of free energy conversion units M^{free} is 2 (c.f. Equation (7.1.2)).

In this chapter, we use the auxiliary problem and present a more general formulation in which an identification of the number of free energy conversion units is not necessary.

7.3 Case studies

The algorithm is applied to two case studies to solve the bilevel problem of scheduling energy and production systems. The production system in the case studies is based on the case study from Kondili et al. (1993) (Figure 3.2), and the model is again based on the State-Task-Network formulation. The energy system model is again based on Voll et al. (2013). In the two case studies, the energy systems differ: In the first case study (CHP subsidies), we consider 3 boilers (4 MW, 1.5 MW, 0.5 MW), and 1 combined-heat-and-power engine (3.5 MW), which are connected to the electricity grid and the gas grid. In the second case study (Grid alternative), we consider: 3 boilers (4 MW, 1.5 MW, 0.5 MW), and 2 combined-heat-and-power engines (1.5 MW, 1.5 MW), which are connected to the electricity grid and the gas grid.

The big-M parameters in Equation (7.22) and Equation (7.29) cannot be calculated exactly. Thus, the values need to be defined big enough to properly model the reformulation as well as small enough to prevent bad numerical behavior. In the case studies, we didn't experience any bad behavior from the choice of our big-M parameters.

For both case studies, we compare the bilevel optimization with the integrated optimization (Chapter 3 and Chapter 4), sequential optimization and the method for incomplete information exchange (Chapter 5 and Chapter 6).

We recall that the integrated optimization of energy and production system optimizes both systems to the objective of the production system (upper-level objective). For the production system, the integrated optimization is the ideal benchmark leading to the lowest cost since the energy system is operated in favor of the production system. The sequential optimization is the most common optimization approach.

If we consider the situation of the production system as the leader and the energy system as the follower, both the integrated and the sequential optimization yield suboptimal solutions for the production system. The solutions are suboptimal because the response of the energy system is too optimistic or not anticipated, respectively. Although the solution of the integrated optimization promises the lowest costs for the production system, the energy system would follow its objective instead of the objective of the production system. To model this behavior, we fix the solution of the integrated optimization for the production system (upper level) variables and optimize the energy system subsequently (lower level). The cost increase for the production system is called regret since it expresses the excess costs that arise if a cooperative energy system is wrongly assumed. In contrast, the bilevel formulation assumes an independent energy system. Thus, the solution of the bilevel formulation provides the minimal realizable cost for the production system.

The algorithm and the benchmarks are executed on an Intel(R) Xeon(R) CPU E5-1660 v3 with 3 GHz running on openSUSE Tumbleweed with Kernelversion 4.17.13. The algorithm is implemented and the subproblems are written in C++ based on an early version of libALE (Djelassi and Mitsos, 2019). The algorithm works in interaction with GAMS such that the subproblems are automatically formulated with their input data as GAMS code. The subproblems are then solved with GAMS 25.1.3 (GAMS Development, 2020) using CPLEX 12.8.0.0 (IBM Corporation, 2020) applying 8 threads. The sequential optimization and the integrated optimization problems are also formulated in GAMS 25.1.3 (GAMS Development, 2020) and solved with CPLEX 12.8.0.0 (IBM Corporation, 2020) applying 8 threads. The time limit for all optimization problems is set to 1000 s, working memory is set to 30 GB, and the allowed absolute gap is 10^{-3}. The optimality tolerance of the upper-level objective function ϵ^{u} is set to 0.01. We choose such tight tolerances since the optimization problems are solved fast and, thereby, we can identify the exact benefits of the bilevel solution compared to the sequential optimization, the integrated optimization and the coordination with incomplete information.

7.3.1 CHP subsidies

In this case study, the energy system gets subsidies for electricity produced by the combined-heat-and-power engines, as it is the case, e.g., in Germany. The subsidies are not forwarded to the production system resulting in misaligned objectives. In Section 7.3.1.1, we describe the detailed setup and objectives. In Section 7.3.1.2, we present and discuss the results of the case study.

7.3.1.1 Formulation

In the bilevel problem for CHP subsidies, the production system is scheduled in the upper-level problem for minimizing its energy and production cost. In the lower-level problem, the operation of the energy system is optimized. In the lower-level problem, the binary variables are: for each energy conversion unit and time step, a binary variable to decide if the energy conversion unit is operated, and a binary variable that allows to only buy electricity from the grid or either sell electricity to the grid at the same time.

The energy system gets subsidies for electricity produced by the combined-heat-and-power engines, but the subsidies are not forwarded to the production system. The amount of subsidies are the German subsidies paid for combined-heat-and-power engines with a capacity over 2 MW. The subsidies and energy prices are provided in Table 7.1. The prices for energy from the grid are not equal in the case studies CHP subsidies and Grid alternative to further distinguish the case studies.

In the following, we present the detailed formulation of the objective functions: The objective function of the production system f^{u} is the production cost $\mathbf{c}^{\mathrm{PS}^T}\mathbf{w}$ and the energy cost to be paid by the production system ($\mathbf{c}^{\mathrm{U,ES,V}^T}\mathbf{V} + \mathbf{c}^{\mathrm{U,ES,o}^T}\mathbf{o}$):

$$
\begin{aligned}
f^{\mathrm{u}}(\mathbf{w},\mathbf{V},\mathbf{o}) &= \mathbf{c}^{\mathrm{PS}^T}\mathbf{w} + \mathbf{c}^{\mathrm{U,ES,V}^T}\mathbf{V} + \mathbf{c}^{\mathrm{U,ES,o}^T}\mathbf{o} \\
&= \sum_{t\in T}\Delta t \cdot \sum_{j\in J}\sum_{i\in I}\sum_{t'=t-\Delta t_{i,j}}^{t}\left(W_{t',i,j}\cdot OC^{\mathrm{fix}}_{i,j} + B_{t',i,j}\cdot OC^{\mathrm{var}}_{i,j}\right) \\
&+ \sum_{t\in T}\Delta t \cdot \sum_{s\in S} V_{t,s}\cdot OC^{\mathrm{stor}}_{s} \\
&+ \sum_{t\in T}\Delta t \cdot \left(p^{\mathrm{buy}}_{gas}\cdot U_{t,gas} + p^{\mathrm{buy}}_{el}\cdot V_{t,u=grid^{\mathrm{buy}},e=\mathrm{el}} - p^{\mathrm{sell}}_{el}\cdot V_{t,u=grid^{\mathrm{sell}},e=\mathrm{el}}\right).
\end{aligned}
\tag{7.35}
$$

The production cost $\mathbf{c}^{\mathrm{PS}^T}\mathbf{w}$ consider fixed cost $OC^{\mathrm{fix}}_{i,j}$ and variable cost $OC^{\mathrm{var}}_{i,j}$ for running task i on production unit j. $W_{t,i,j}$ is a binary variable and equals 1 only in time step t when task i started on production unit j. $B_{t,i,j}$ equals the batch size only in time step t when task i started on production unit j. Furthermore, variable cost OC^{stor}_{s} is considered for storing amount $V_{t,s}$ of product s in time step t.

The energy cost $\mathbf{c}^{\mathrm{U,ES,V}^T}\mathbf{V} + \mathbf{c}^{\mathrm{U,ES,o}^T}\mathbf{o}$ considers cost for purchasing gas, purchasing electricity, and revenues for selling electricity. The cost for electricity results from purchasing electricity $V_{t,u=grid^{\mathrm{buy}},e=\mathrm{el}}$ for a price of $p^{\mathrm{el,buy}}$. The revenues from selling electricity result from selling electricity $V_{t,u=grid^{\mathrm{sell}},e=\mathrm{el}}$ for a price of p^{sell}_{el}. The cost for gas results from purchasing gas $U_{t,gas}$ for the price of p^{buy}_{gas}. The amount of consumed gas is calculated by an affine function depending on $V_{t,u,e=\mathrm{heat}}$ and $o_{t,u}$. For the cost to be paid to the energy system, a profit margin of the energy system could be added.

Table 7.1: Energy cost and subsidies of case study CHP subsidies. $p_{el}^{\text{CHP,sell}}$ and $p_{el}^{\text{CHP,prod}}$ are subsidies for electricity from combined-heat-and-power engines sold to the grid or consumed on-site, respectively. p_{gas}^{buy} is the gas price. p_{el}^{buy} and p_{el}^{sell} is the electricity price for purchase and selling, respectively.

energy	$p_{el}^{\text{CHP,sell}}$	$p_{el}^{\text{CHP,prod}}$	p_{gas}^{buy}	p_{el}^{buy}	p_{el}^{sell}
prices or subsidies /(€/kWh)	0.031	0.018	0.05	0.04	0.035

Here, we assume only a profit margin for the energy system by subsidies for running combined-heat-and-power engines.

The objective function of the energy system is the cost to be paid by the energy system:

$$\begin{aligned} f^{\text{l}}(\mathbf{V}, \mathbf{o}) &= \mathbf{c}^{\text{L,ES,V}^T}\mathbf{V} + \mathbf{c}^{\text{L,ES,o}^T}\mathbf{o} \\ &= \sum_{t \in T} \Delta t \cdot \Bigg[p_{gas}^{\text{buy}} \cdot U_{t,gas} + p_{el}^{\text{buy}} \cdot V_{t,u=grid^{\text{buy}},e=\text{el}} \\ &- \left(p_{el}^{\text{sell}} + p_{el}^{\text{CHP,sell}} \right) \cdot V_{t,u=grid^{\text{sell}},e=\text{el}} \\ &- p_{el}^{\text{CHP,prod}} \cdot \left(E_{t,e=\text{el}}^{\text{demand}} - V_{t,u=grid^{\text{buy}},e=\text{el}} \right) \Bigg]. \end{aligned} \tag{7.36}$$

These costs result from purchasing gas and electricity, revenues for selling electricity, and subsidies for running combined-heat-and-power engines. The cost for gas is again calculated by the amount of gas $U_{t,gas}$ and the price of gas p_{gas}^{buy}. The cost for electricity is calculated by the bought electricity $V_{t,u=grid^{\text{buy}},e=\text{el}}$ and the price for electricity $p^{\text{el,buy}}$. The revenues for selling electricity result from selling electricity $V_{t,u=grid^{\text{sell}},e=\text{el}}$ for a price of p_{el}^{sell}. The revenues for selling electricity are increased by the subsidies $p_{el}^{\text{CHP,sell}}$ for electricity produced by the combined-heat-and-power engines. Further subsidies $p_{el}^{\text{CHP,prod}}$ are gathered for electricity produced by the combined-heat-and-power engines and not sold to the grid.

The case study CHP subsidies considers a time horizon of 12 h. Each time step has the same length Δt of 1 h. In the last time step of the time horizon, the production system has to supply 56 t of S7 and 108 t of S10. The parameters for equipment size, storage size, and operational cost of the production system are the same as reported in Chapter 5. The parameters of the energy conversion units are taken again from Voll et al. (2013).

7.3.1.2 Results

The optimal bilevel solution ($f^{\mathrm{u,bilvl}} = 4936$ €) saves 3.3 % of the cost for the production system compared to the sequential optimization ($f^{\mathrm{u,seq}} = 5106$ €, c.f. Figure 7.1). The method from Chapter 5 results in $f^{\mathrm{u,inc.inf.}} = 5077$ € cost for the production system and saves 0.6 % compared to the sequential optimization. The integrated optimization results in even lower cost ($f^{\mathrm{u,int}} = 4637$ €) than the bilevel optimization and saves 9.2 % compared to sequential optimization. However, as mentioned previously, the integrated optimization assumes full cooperation of the energy system, which is not given in this case study. If we subsequently optimize the energy system to calculate the regret, the cost for the energy system decreases, but cost increases for the production system ($f^{\mathrm{u,corr.\ int}} = 5117$ €). The cost increase for the production system (regret) is 9.4 %. The regret results in even higher cost for the production system than in the sequential optimization (0.2 %). Thus, the bilevel optimization provides the minimal, realizable cost for the production system (Figure 7.1).

Substantial cost benefits arise in the integrated optimization because the combined-heat-and-power engines are not operated (Figure 7.2). Thus, no heat and electricity is supplied by the combined-heat-and-power engines and instead by the boilers and the electricity grid. Electricity from the electricity grid is cheaper for the production system, but for the energy system, the subsidies make the combined-heat-and-power engines more beneficial. Thus, the production system favors that the energy system supplies electricity from the grid and the energy system favors to supply the electricity by the combined-heat-and-power engines. In the sequential optimization, 45.9 % of the electricity is covered by the combined-heat-and-power engines. In hours where the energy system can produce more electricity than demanded, the energy system sells additional electricity to the grid.

In the bilevel optimization, 20.1 % of the electricity demand is covered by combined-heat-and-power engines because the energy system significantly benefits when using combined-heat-and-power engines. The production system chooses a demand profile which results in low supply of electricity by the combined-heat-and-power engines due to the relatively high cost of electricity covered by combined-heat-and-power engines.

The subproblems of the bilevel algorithm, the method from Chapter 5, the integrated optimization, and the sequential optimization are solved within the time limit and, thus, are solved to the predefined gap. The sequential optimization and integrated optimization are solved each within 2 s. The method from Chapter 5 solves the problem in 23 s. The proposed algorithm solves the bilevel optimization problem in 262 s and needs 4 iterations. The lower-level problem and the auxiliary problem

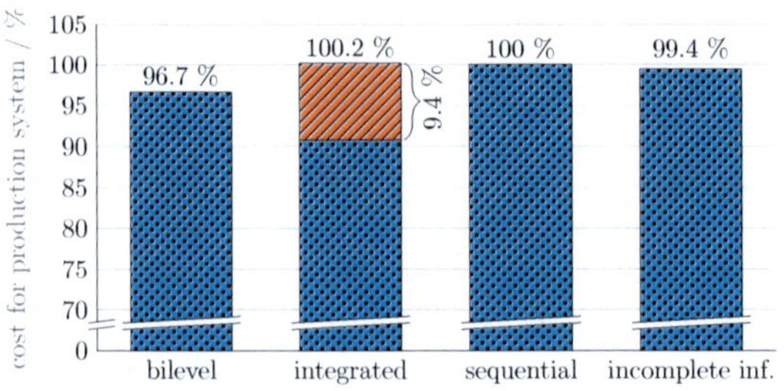

Figure 7.1: Cost from the bilevel and the alternative optimization approaches in the case study CHP subsidies. In the integrated optimization, the cost of the production system is minimized in a single-level optimization. The cost of regret is added after a subsequent optimization of the energy system. In the sequential optimization, first, the production system is optimized without considering energy cost. Subsequently, the energy system is optimized. The method from Chapter 5 uses demand-dependent energy cost as incomplete information (incomplete inf.).

are always solved well below 1 s. The solution time of the lower-bounding problem increases with each iteration with <1 s, 12 s, 72 s, and 178 s. The solution time increases since discretization points are added in each iteration.

The algorithm identifies 13 discretization points in total. From the first iteration, 7 discretization points are identified, the second iteration identifies 5 additional discretization points, and finally, in the third iteration, 1 additional discretization point is identified. With these 13 discretization points in the fourth iteration, the lower-bounding problem reaches the same result as the evaluation of the lower-level solution in the upper-level objective, and the algorithm terminates with the optimal bilevel solution.

As expected, the costs resulting from the lower bound increase in each iteration (Figure 7.3) because in each iteration, we add tightening constraints by the discretization points. No such trend exists for the optimal solution of the sum of the cost from the lower-bounding problem and the additional cost from evaluation of the lower-level solution in the upper-level objective. Similar to the integrated optimization, these additional costs are named regret because they also arise from a subsequent optimization of the energy system. Since the lowest cost can only be reached by the optimal

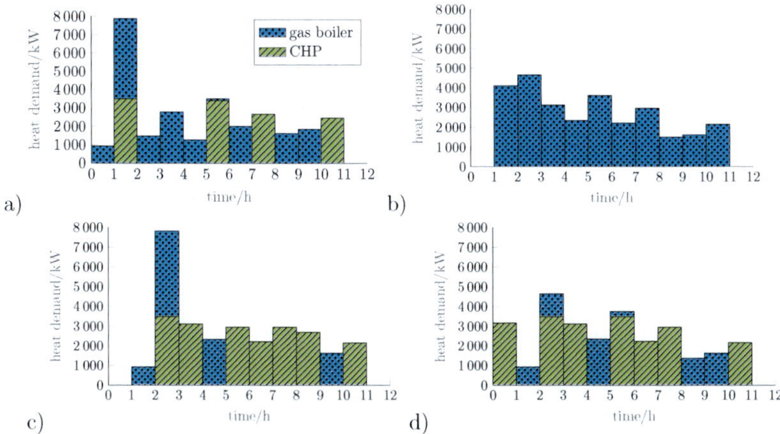

Figure 7.2: Heat demands in the case study CHP subsidies. The heat demands are plotted for: a) bilevel optimization b) integrated optimization, c) sequential optimization and d) the method using incomplete information.

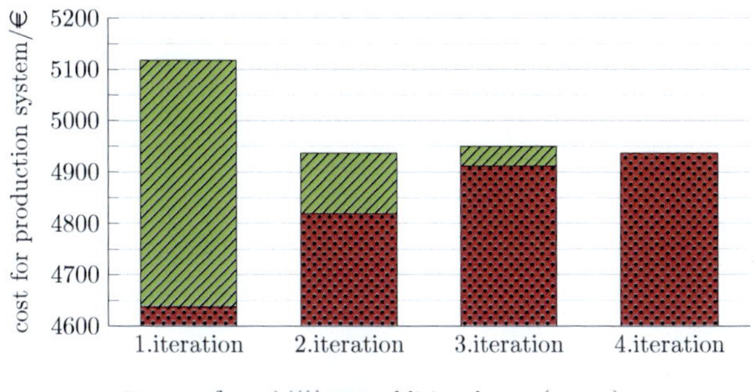

Figure 7.3: Total cost for the production system in the case study CHP subsidies from the lower bound (LBD) and additional cost by evaluation of the lower-level solution in the upper-level objective (regret). The cost for the production system from the lower bound (LBD) increases with each iteration.

solution of the bilevel problem, the costs from the lower-bounding problem and the regret are never lower than the optimal solution of the bilevel problem (Figure 7.3).

7.3.2 Grid alternative

In this case study, the energy system is only connected to the gas grid. Thus, the energy system provides heat and electricity, while it provides electricity only by combined-heat-and-power engines. The production system can buy additional electricity directly from the electricity grid. First, we describe the detailed setup and the objectives (Section 7.3.2.1). Second, we present the results (Section 7.3.2.2).

7.3.2.1 Formulation

In the bilevel problem of Grid alternative, the production system is scheduled in the upper level to minimize its energy and production cost. The energy system is scheduled in the lower level. In the lower-level problem, we have the binary variables as in the case study CHP subsidies (Section 7.3.1.1).

In this case study, the production system pays a predefined price for heat and electricity to the energy system. Additionally, the production system is directly connected to the grid and can purchase electricity from the grid. Thus, the energy system has to cover the heat demand but not the electricity demand. The energy prices are given in Table 7.2. The objective function of the upper level (production system) is $f^{\mathrm{u}}(\mathbf{w}, \mathbf{V}, \mathbf{o})$ and considers production cost ${\mathbf{c}^{\mathrm{PS}}}^T\mathbf{w}$ and energy cost $({\mathbf{c}^{\mathrm{U,ES,V}}}^T\mathbf{V} + {\mathbf{c}^{\mathrm{U,ES,o}}}^T\mathbf{o})$ to be paid by the production system:

$$\begin{aligned} f^{\mathrm{u}}(\mathbf{w}, \mathbf{V}, \mathbf{o}) =& {\mathbf{c}^{\mathrm{PS}}}^T\mathbf{w} + {\mathbf{c}^{\mathrm{U,ES,V}}}^T\mathbf{V} + {\mathbf{c}^{\mathrm{U,ES,o}}}^T\mathbf{o} \\ =& {\mathbf{c}^{\mathrm{PS}}}^T\mathbf{w} + \sum_{t\in T} \Delta t \cdot \Big[p_{heat}^{\mathrm{buy}} \cdot E_{t,e=\mathrm{heat}}^{\mathrm{demand}} \\ & + p_{el}^{\mathrm{buy,ES}} \cdot \left(E_{t,e=\mathrm{el}}^{\mathrm{demand}} - V_{t,u=grid^{\mathrm{buy}},e=\mathrm{el}} \right) + p_{el}^{\mathrm{buy,grid}} \cdot V_{t,u=grid^{\mathrm{buy}},e=\mathrm{el}} \Big]. \end{aligned} \tag{7.37}$$

The production cost ${\mathbf{c}^{\mathrm{PS}}}^T\mathbf{w}$ is the same as in case study CHP subsidies (Equation (7.35)). The energy cost ${\mathbf{c}^{\mathrm{U,ES,V}}}^T\mathbf{V} + {\mathbf{c}^{\mathrm{U,ES,o}}}^T\mathbf{o}$ consider cost for purchasing heat and electricity from the energy system as well as cost for electricity purchased directly from the grid. The cost for heat is calculated by the heat demand $E_{t,e=\mathrm{heat}}^{\mathrm{demand}}$, and the price for heat p_{heat}^{buy}. The cost for electricity from the energy system is calculated by the price for electricity from the energy system $p_{el}^{\mathrm{buy,ES}}$ multiplied by the difference of electricity demand $E_{t,e=\mathrm{el}}^{\mathrm{demand}}$ and electricity already bought from the grid $V_{t,u=grid^{\mathrm{buy}},e=\mathrm{el}}$. The cost for electricity from the grid is calculated by the price for electricity from the grid $p_{el}^{\mathrm{buy,grid}}$, and the amount of electricity bought from the grid $V_{t,u=grid^{\mathrm{buy}},e=\mathrm{el}}$.

Table 7.2: Energy price of case study Grid alternative. p_{heat}^{buy} and $p_{el}^{\text{buy,ES}}$ are the prices for heat and electricity to be paid by the production system to the energy system, respectively. $p_{el}^{\text{buy,grid}}$ is the price for the production system for electricity from the grid. p_{gas}^{buy} is the gas price and p_{el}^{sell} is the electricity price for the energy system.

energy	p_{heat}^{buy}	$p_{el}^{\text{buy,ES}}$	$p_{el}^{\text{buy,grid}}$	p_{gas}^{buy}	p_{el}^{sell}
price /(€/kWh)	0.07	0.05	0.2	0.06	0.04

The objective function of the lower level (energy system) is to maximize the profit (equal to minimization of negative profit). The profit of the energy system $P^{\text{L,ES}}$ considers revenues from selling heat and electricity to the production system, cost for purchasing gas and revenues for selling electricity to the grid:

$$
\begin{aligned}
f^{\text{l}}(\mathbf{V}, \mathbf{o}) =& \mathbf{c}^{\text{L,ES,V}^T}\mathbf{V} + \mathbf{c}^{\text{L,ES,o}^T}\mathbf{o} = -P^{\text{L,ES}} \\
=& -\sum_{t \in T} \Delta t \cdot \Big[p_{heat}^{\text{buy}} \cdot E_{t,e=\text{heat}}^{\text{demand}} + p_{el}^{\text{buy,ES}} \cdot \left(E_{t,e=\text{el}}^{\text{demand}} - V_{t,u=grid^{\text{buy}},e=\text{el}} \right) \\
& - p_{gas}^{\text{buy}} \cdot U_{t,gas} + p_{el}^{\text{sell}} \cdot V_{t,u=grid^{\text{sell}},e=\text{el}} \Big].
\end{aligned}
\tag{7.38}
$$

The revenues from heat are calculated by the heat demand $E_{t,e=\text{heat}}^{\text{demand}}$ and the price for heat p_{heat}^{buy}. The revenues from selling electricity to the production system are calculated by the price for electricity $p_{el}^{\text{buy,ES}}$ multiplied by the difference between electricity demand $E_{t,e=\text{el}}^{\text{demand}}$ and electricity already bought from the grid $V_{t,u=grid^{\text{buy}},e=\text{el}}$. The cost of purchasing gas results from purchasing amount of gas $U_{t,gas}$ to run the energy conversion units, and the price for gas p_{gas}^{buy}. The amount of consumed gas is calculated by an affine function depending on $V_{t,u,e=\text{heat}}$ and $o_{t,u}$. The revenues from selling electricity result from selling amount of electricity $V_{t,u=grid^{\text{buy}},e=\text{el}}$ for a price of p_{el}^{sell}.

In the last time step of the time horizon, the production system has to supply the same amount as in the case study CHP subsidies, i.e., 56 t of S7 and 108 t of S10. All other parameters are the same as in the case study CHP subsidies.

7.3.2.2 Results

Again, the bilevel optimization provides the minimal realizable cost for the production system ($f^{\text{u,bilvl}} = 6689\,€$).

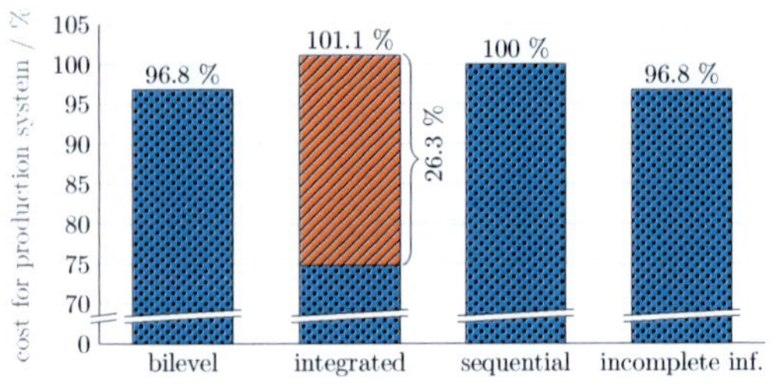

Figure 7.4: Cost from the bilevel and alternative optimization approaches in the case study Grid alternative. In the integrated optimization, the cost of the production system is minimized in a single-level optimization. The cost of regret is added after a subsequent optimization of the energy system. In the sequential optimization, first, the production system is optimized without considering energy cost. Subsequently, the energy system is optimized. In this case study, the method from Chapter 5 using demand-dependent energy cost as incomplete information reaches the cost from the bilevel optimization (incomplete inf.).

The bilevel optimization saves 3.2 % compared to sequential optimization ($f^{\mathrm{u,seq}} = 6910$ €, Figure 7.4). In this case study, the method from Chapter 5 reaches the same cost ($f^{\mathrm{u,inc.inf.}} = 6689$ €) as the bilevel optimization. Still, as shown in the previous case study CHP subsidies, the identification of the bilevel solution is not guaranteed. The integrated optimization ($f^{\mathrm{u,int}} = 5170$ €) would result in 25.2 % cost savings for the production system compared to the sequential optimization. As in the case study CHP subsidies, the integrated optimization results in the best objective for the production system, but the solution from the integrated optimization is again not applicable in practice. The cost for the integrated optimization is increased after a subsequent lower-level optimization ($f^{\mathrm{u,corr.\ int}} = 6989$ €). The regret is 26.3 %, and consequently, the cost for the production system is even 1.1 % higher than from the sequential optimization.

The cost benefits of the integrated optimization arise from energy supply by the combined-heat-and-power engines (Figure 7.5). The combined-heat-and-power engines are operated such that electricity demand is totally supplied by the combined-heat-and-power engines because, for the production system, electricity from the energy system is cheaper than electricity from the grid. In the sequential optimization and

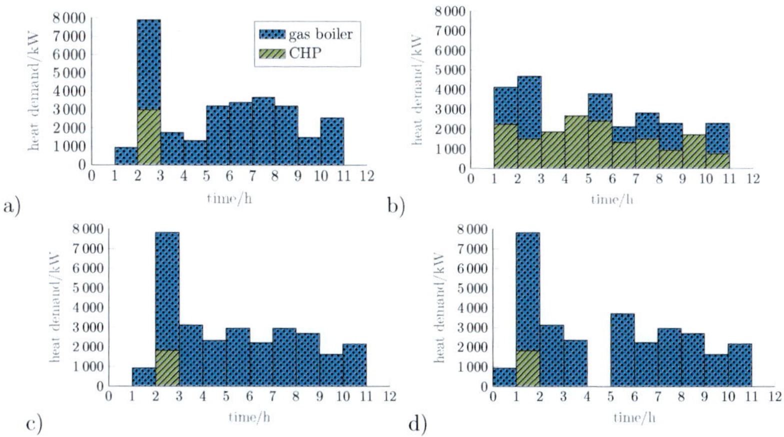

Figure 7.5: Heat demands in the case study Grid alternative. The heat demands are plotted for: a) bilevel optimization b) integrated optimization, c) sequential optimization and d) the method using incomplete information.

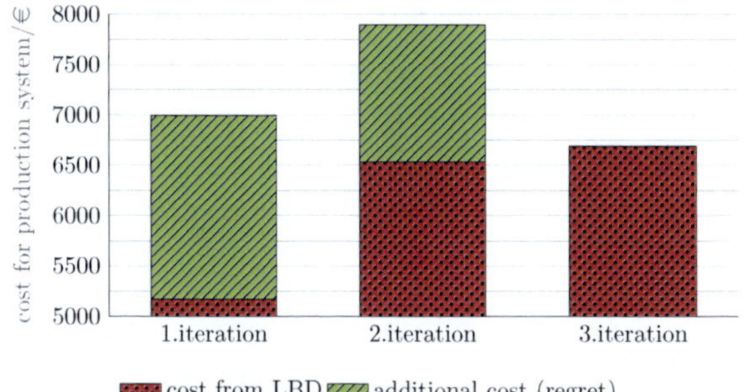

Figure 7.6: Total cost for the production system in the case study Grid alternative from the lower bound (LBD) and from the evaluation of the lower-level solution in the upper-level objective (regret). The costs for the production system from the lower bound (LBD) increase with each iteration.

in the bilevel optimization, only 4 % and 4.8 % of electricity demand is covered by combined-heat-and-power engines, respectively. This low electricity coverage by the combined-heat-and-power engines is caused because the energy system does not prefer to use combined-heat-and-power engines to cover the electricity demand since the revenues for electricity are low. This difference in the use of the combined-heat-and-power engines results in the large cost differences between the integrated optimization and the other approaches.

The sequential and integrated optimizations are solved within 2 s. The method from Chapter 5 solves the problem within 30 s. The proposed algorithm solves the bilevel optimization problem in 153 s and needs 3 iterations. The lower-level problem and the auxiliary problem are always solved well below 1 s. The solution time of the lower-bounding problem increases with each iteration with <1 s, 84 s, and 68 s. The algorithm identifies 7 discretization points. Thus, fewer discretization points than time steps are needed, meaning that some discretization points are valid and used for multiple time steps. In the first iteration, 4 discretization points are identified, and in the second iteration, 3 discretization points are identified. The algorithm terminates in the third iteration, and the optimal solution of the bilevel problem is identified.

As expected, the cost resulting from the lower-bounding problem increase in each iteration (Figure 7.6), and the cost resulting from the evaluation of the lower-level solution in the upper-level objective does not show a trend.

The case studies show that the proposed algorithm solves the bilevel problem efficiently with only a few iterations. Furthermore, the identified bilevel solution considers the misaligned objectives of energy and production system and if complete information on the energy system is available, the bilevel solution allows for the best solution in practical application.

7.4 Conclusions

In this Chapter, we formulate a bilevel problem for energy and production system scheduling. To solve this bilevel problem, we select the relevant parts from the algorithm in Djelassi et al. (2019) and add a procedure to identify dependent and independent variables. Thereby, we can solve the bilevel problem of energy and production system scheduling.

The algorithm iteratively solves the bilevel problem, while 3 optimization problems are solved in each iteration: lower-bounding problem, lower-level problem, and auxiliary problem. In the lower-bounding problem, discretization points bound the lower-level objective function by solutions of the lower-level problem from previous iterations.

The algorithm is successfully applied to two case studies. The solution of the bilevel problem reaches cost savings of 3.3 % and 3.2 % compared to the sequential optimization, and 3.5 % and 4.3 % compared to an integrated optimization including regret. Furthermore, in the first case study, cost savings are realized compared to the method proposed in Chapter 5. In the second case study, the bilevel optimization and the method proposed in Chapter 5 reach the same objective function.

Thus, on the one hand, the proposed algorithm solves the bilevel problem. On the other hand, the solution of the bilevel problem provides the best realizable solution in a Stackelberg game where the production system is the leader and the energy system is the follower.

Chapter 8

Summary, conclusions and future perspectives

In this chapter, we first give an overall summary and draw conclusions from this thesis (Section 8.1). Next, we present future perspectives on the optimization methods presented in this thesis (Section 8.2). Finally, future research topics are discussed to improve the optimization of integrated industrial sites (Section 8.3).

8.1 Summary and conclusions

The industrial sector plays a key role in the reduction of greenhouse gas emissions. The majority of the greenhouse gas emissions in the industrial sector are energy-related emissions (IPCC, 2014). Thus, to reach climate goals, improvements in the energy supply of the industrial sector are necessary.

Commonly, industrial sites consist of energy and production systems. Typically, the systems are designed and scheduled individually, which leads to increased overall costs and less efficient energy supply and use.

In this thesis, optimization methods are proposed for the optimization of integrated energy and production systems. For the optimization methods, we distinguish two cases for the relationship between energy and production systems. In one case, both systems follow the same objective, and in the other case, the systems follow misaligned objectives. For the case in which both systems follow the same objective, we present two methods for the integrated optimization of energy and production system. In the first method, presented in Chapter 3, we propose the integrated synthesis of energy and production systems. The systems are coupled by the energy demand of the production system, which has to be supplied by the energy system. The integration of both optimization problems leads to an overall superstructure, which allows optimizing the overall system by exploiting design trade-offs between the production system and the energy system.

In the second method for the integrated optimization (Chapter 4), we propose a method for an integrated scheduling of energy and production systems for control-reserve provision. The method optimizes the production schedule and, consequently, its energy demand for provision of control reserve by the energy system. The method employs a two-stage stochastic optimization, which optimizes the schedule of the energy system if control reserve is requested. Thus, only the schedule of the energy system is adapted if control reserve is requested. The method enables energy and production system operators to schedule integrated systems for participation in control-reserve markets.

For the case, in which both systems follow misaligned objectives, e.g., two different companies operate the systems, the integrated optimization to a single objective is not suitable. We describe the relationship between energy and production system as a Stackelberg game. The Stackelberg game is further classified if the systems have incomplete or complete information on each other. In Chapter 5 and Chapter 6, we present methods if only incomplete information is exchanged between single or multiple energy and production systems, respectively. For the incomplete information exchange between a single energy and production system, a method is proposed in Chapter 5 based on a repeated Stackelberg game. In the Stackelberg game, demand-dependent energy costs are introduced to pass incomplete information on the costs of energy supply from the energy system to the production system. The demand-dependent energy costs are based on the assumption that the energy cost structure of the energy system can be sufficiently approximated by a piecewise-affine function. In Chapter 6, we extend the proposed method to multiple energy and production systems. For this purpose, we propose novel coordination methods among the production systems and among the energy systems. In particular, we propose a method to allocate the demand-dependent energy cost to the production systems. Furthermore, since the energy supplied by each energy systems is decided by the energy price, we propose an optimization problem for overall energy-price minimization.

In Chapter 7, we consider complete information exchange between energy and production system. For this case, we formulate a bilevel problem for energy and production system scheduling. For the solution of the bilevel problem, we propose an algorithm that iteratively solves the bilevel problem. Within each iteration, three optimization problems are solved: lower-bounding problem, lower-level problem, and auxiliary problem. The bilevel solution is found in the lower-bounding problem where discretization points from lower-level problem solutions are used to bound the lower-level objective.

In each of the Chapters 3 to 7, the methods are applied to case studies presenting the benefits of the proposed methods compared to the commonly used approaches for optimization. Finally, in Chapter 7, we apply the common sequential optimization, the integrated optimization, the proposed method using incomplete information, and the algorithm to solve the bilevel problem to the same case study (Section 7.3). The comparison shows that an integrated optimization is desirable if all systems follow the same objective. However, an integrated optimization is not suitable for the Stackelberg game. If only incomplete information is available, the proposed method using the concept of demand-dependent energy cost already provides good solutions. However, the optimal solution for the Stackelberg game is not guaranteed. If complete information is available in the Stackelberg game, a bilevel problem can be formulated and solved. The optimal solution of the bilevel problem is the optimal solution of the Stackelberg game.

The case studies show that a single optimization method cannot provide the optimal solution for all relationships between energy and production systems. Methods have to be tailored to the individual situation. This thesis presents optimization methods for all the identified relationships such that system operators of energy and production systems can choose a method for their situation to yield the optimal result.

8.2 Future perspectives on the optimization methods

In the following, we present possible future improvements and extensions of the optimization methods presented in this thesis. These improvements result from a critical analysis of the current status of the methods and the possible extensions are based on the current methods. Thus, the suggestions are made for the individual methods. The improvements and extensions are ordered by increasing expected effort.

Time coupling. The methods in Part II to optimize the Stackelberg game do not account for time coupling in the energy systems, e.g., by storage units. Thus, time-coupling effects are not considered and units such as storage units are not considered. To consider time coupling in the energy system, the concept of demand-dependent energy cost needs to be extended by the dimension of time in Chapter 5 and Chapter 6. In Chapter 7, the discretization points to represent the solution of the energy system are defined for the energy demand in a single time step. These discretization points need to be extended such that the optimal solution is represented for multiple time steps, i.e., an energy demand curve. The consideration of time coupling allows considering energy conversion units such as storage units or time-coupling effects such as start-up times.

Uncertainty in production systems. In Chapter 4, a method is proposed for control-reserve provision by an integrated energy and production system. In this method, the energy system reacts to the request of control reserve. However, the production schedule is fixed. The method can be further extended by reactive scheduling of the production system. In reactive scheduling, the production schedule is quickly adapted if an unforeseen event occurs (Janak et al., 2006). An unforeseen event is the request of control reserve. Thus, the production schedule is adapted and rescheduled if control reserve is requested. Hence, the production system can adjust its energy demand as a reaction to control-reserve requests. This adjustments would allow for an increased provision of control reserve by the integrated system.

Bilevel optimization considering multi-leader and multi-follower. The bilevel optimization algorithm considers one upper and one lower level. However, as we already considered in Chapter 6, there can also be multiple players in the upper level (production systems) and multiple players in the lower level (energy systems) on the same industrial site. If a Nash game exists on each level, i.e., a Nash game between the multiple leader and a Nash game between the multiple follower, a Nash equilibrium needs to be found for each level. Finding these Nash equilibria without integer decisions is already challenging, but solution procedures have already been proposed for eco-industrial parks (Ramos et al., 2018).

8.3 Future perspectives on scope expansion

In the following, we propose future scope expansion which are related to the content of this thesis. We discuss how the scope considered in this thesis can be expanded to other domains. Thus, we extend the motivation of integration beyond energy and production systems. The expansions are ordered by increasing expected effort.

Total Site Heat Integration. In this thesis, the energy system supplies energy without considering the reuse of heat. However, if the temperatures of cooling and heating demands are within the same range, heat integration can be performed. For example, suppose a process with a high temperature level needs to be cooled, and there is a heating demand at a lower temperature level. In this case, both energy demands can be integrated. Considering the heat integration of all systems in an industrial site is commonly called Total Site Heat Integration (Liew et al., 2017). The optimization methods proposed in this thesis could be extended to the research field of heat integration, and thus, account for a Total Site Heat Integration. This Total Site Heat Integration increases the efficiency of energy supply and use in the industrial site.

Consideration of multiple levels. In this thesis, we considered two levels in the Stackelberg game. Often, more than two levels are involved in a decision process. For example, the procurement of secondary energy for the energy system is sometimes considered as an individual player, or the consumer of the produced goods could be represented as a market at the uppermost level. The methods for incomplete information and bilevel optimization could be extended to solve a multi-level Stackelberg game. For the incomplete information exchange, new or adapted concepts need to be developed. The bilevel optimization algorithm needs to be extended such that discretization points for the uppermost level represent the optimal solution of the second uppermost level, which is already a solution of a bilevel problem. The algorithm will become further complicated by coupling constraints between all levels. However, the consideration of multiple levels allows for the representation of more complex situations.

Integrated and Stackelberg-based optimization in other domains. The optimization methods presented in this thesis are all developed for industrial sites. However, multiple players interact with each other also in other domains. Thus, an integrated optimization, a coordination of the optimization problems, or a bilevel optimization might be applicable. Possible domains where the proposed methods could be applied are supply-chain planning (Yeh et al., 2015) and energy markets (Wogrin et al., 2020). Wogrin et al. (2020) point out in their review on bilevel problems in energy markets that improvement is needed to consider binary variables in the lower-level problem. The Stackelberg game-based methods proposed in this thesis consider binary variables, and thus, could be adapted and applied for these problems.

Multiple energy markets. In Chapter 4, we extended the integrated scheduling for participation in the tertiary control-reserve market. The integrated scheduling could further be extended to consider multiple energy markets. Short-term electricity markets are the day-ahead electricity market and the intra-day electricity market. Other markets for control-reserve provision are the primary and secondary control-reserve markets. However, prices are uncertain in all these markets and in some of them, even the delivery of energy is uncertain. Thus, considering multiple markets in the integrated optimization would result in a multi-stage stochastic program. We expect the solution of the multi-stage stochastic program as computationally challenging such that decomposition methods and scenario reduction techniques are necessary. However, if all markets are considered in one optimization, the optimal allocation of energy purchase and sale as well as the optimal allocation of flexibility can be reached.

In this thesis, we have proposed optimization methods for integrated energy and production systems to unlock synergies and have overcome challenges resulting from the relationships between energy and production systems. Ultimately, the examples for future perspectives show that much potential for integration of energy and production systems remains beyond the scope of this thesis. More precisely, the future research topics show a need for integration of Energy and Process Systems Engineering to other domains. Thus, this thesis is only one step towards an overall integration that will shape a more efficient industry with fewer greenhouse gas emissions. Of course, during this transition, many new and unforeseen problems will arise and need to be solved, but

'A problem is a chance for you to do your best.'

Duke Ellington

Appendices

Appendix A

Optimization models of energy and batch production systems

In this chapter, we present the optimization models of the energy and production system. Here, the basic constraints for scheduling and synthesis are given. The additional constraints necessary for the method from Chapter 5 are given in Appendix A.2.

Major parts of this chapter are reproduced by permission of Elsevier with modifications from the complete Elsevier from:

Leenders, L., Bahl, B., Hennen, M., and Bardow, A. (2019a). Coordinating scheduling of production and utility system using a Stackelberg game. *Energy*, 175, 1283-1295.

Leenders, L., Bahl, B., Lampe, M., Hennen, M., and Bardow, A. (2019b). Optimal design of integrated batch production and utility systems. *Computers & Chemical Engineering*, 128, 496-511.

The author of this thesis contributed to the development and implementation of the methods as well as the calculation and interpretation of the results, wrote the draft of the papers and is their principal author.

A.1 Model formulation for the synthesis and scheduling of energy and production systems

In this section, we briefly summarize the model formulation for the synthesis and scheduling of energy and production systems. For only scheduling the systems, the design decisions are fixed.

A.1.1 Production system

The production system is a batch production system, and the modeling is based on the work of Pinto et al. (2008).

Only if production equipment j is installed ($\gamma_j = 1$), a task can be processed on a production equipment j in time step t:

$$\gamma_j \geq \sum_{i \in I} \sum_{t'=t-\Delta t_{i,j}^{\text{process}}+1}^{t} W_{t',i,j} \quad \forall t \in T, j \in J. \tag{A.1}$$

The binary variable $W_{t,i,j}$ identifies if task i has started on production unit j in time step t.

If production equipment j is installed, the size of the production equipment V_j equals the size of the greatest batch $B_{t,i,j}$ processed on the corresponding production equipment. Since greater equipment increases the objective function, the following inequality constraint provides an adequate formulation:

$$V_j \geq B_{t,i,j} \quad \forall t \in T, j \in J, i \in I. \tag{A.2}$$

The size of an installed production equipment V_j is constrained by the lower and upper bound (V_j^{lb} or V_j^{ub}) of the production equipment size:

$$\gamma_j \cdot \mathrm{V}_j^{lb} \leq V_j \leq \gamma_j \cdot \mathrm{V}_j^{ub} \quad \forall j \in J. \tag{A.3}$$

If a storage unit s is installed, the size of the storage unit V_s equals the size of the greatest storage level $V_{t,s}$:

$$V_s \geq V_{t,s} \quad \forall t \in T, s \in S. \tag{A.4}$$

The size of an installed storage unit is bound by the lower and upper bound (V_s^{lb} or V_s^{ub}) of the storage unit size:

$$\gamma_s \cdot \mathrm{V}_s^{lb} \leq V_s \leq \gamma_s \cdot \mathrm{V}_s^{ub} \quad \forall s \in S. \tag{A.5}$$

Piping is considered between a storage unit s and production equipment j. If a storage unit is installed, a piping unit is installed. The size of the piping unit $V_{s,j}^{\text{piping}}$ equals the size of the greatest batch $B_{t,i,j}$ that is pumped through the pipe. Pipes are installed before and after a storage. $\mathrm{b}_{s,i}^{\text{con. frac}}$ and $\mathrm{b}_{s,i}^{\text{prod. frac}}$ are the consumption and production fraction, respectively, that task i needs from storage s to run the batch $B_{t,i,j}$:

$$V_{s,j}^{\text{piping}} \geq \sum_{i \in I} (\mathrm{b}_{s,i}^{\text{con. frac}} \cdot B_{t,i,j}) \;\; \forall t \in T, s \in S, j \in J \tag{A.6}$$

$$V_{s,j}^{\text{piping}} \geq \sum_{i \in I} (\mathrm{b}_{s,i}^{\text{prod. frac}} \cdot B_{t,i,j}) \;\; \forall t \in T, s \in S, j \in J. \tag{A.7}$$

The size of an installed piping unit is bound by the lower and upper bound ($\mathrm{V}_{s,j}^{lb}$ and $\mathrm{V}_{s,j}^{ub}$) of the piping unit size:

$$\gamma_{s,j} \cdot \mathrm{V}_{s,j}^{lb} \leq V_{s,j}^{\text{piping}} \leq \gamma_{s,j} \cdot \mathrm{V}_{s,j}^{ub} \;\; \forall s \in S, j \in J. \tag{A.8}$$

Equation (A.9) guarantees the satisfaction of the product demand:

$$V_{t,s} = V_{t-1,s} + \sum_{i \in I} \sum_{j \in J} (\mathrm{b}^{\text{prod. frac}} \cdot B_{t-\Delta \mathrm{t}_{i,j}^{\text{process}},i,j}) - \sum_{i \in I} \sum_{j \in J} (\mathrm{b}^{\text{con. frac}} \cdot B_{t,i,j})$$
$$+ \text{raw}_{t,s} - \text{product}_{t,s} \;\; \forall t \in T, s \in S. \tag{A.9}$$

The storage level $V_{t,s}$ increases and decreases by incoming and outgoing batches and is reduced by the produced product $\text{product}_{t,s}$ in the corresponding time step. $\text{raw}_{t,s}$ is the incoming raw material.

The energy demand is calculated as explained in Equation (3.5) and contains a variable and a fixed energy demand. If the production system already exists, the size of the equipment is fixed, i.e., production equipment size V_j, storage unit size V_s, piping unit size $V_{s,j}^{piping}$.

A.1.2 Energy system

The model of the energy system is based on the work of Voll et al. (2013). The calculation of the annual operational expenditure of the energy system $OPEX^{\text{ES}}$ is given in general form in Equation (3.8). Here, we specify the equation for an energy system running on electricity and gas:

$$OPEX^{\text{ES}} = \sum_{u \in U} CM_u$$
$$+ \frac{\mathrm{t}^{\text{year}}}{\mathrm{t}^{\text{cycle}}} \cdot \sum_{t \in T} \left[\Delta t \cdot \left(P_{t,el}^{\text{buy}} \cdot \mathrm{p}_{el}^{\text{buy}} - P_{t,el}^{\text{sell}} \cdot \mathrm{p}_{el}^{\text{sell}} + \sum_{u \in U} (U_{t,u}^{\text{gas}} \cdot \mathrm{p}_{gas}^{\text{buy}}) \right) \right]. \tag{A.10}$$

The $OPEX^{\text{ES}}$ takes into account costs for maintenance CM_u, costs for purchased electricity $P_{t,el}^{\text{buy}}$ at a price of $\mathrm{p}_{el}^{\text{buy}}$, costs for purchased gas $U_{t,u}^{\text{gas}}$ at a price of $\mathrm{p}_{gas}^{\text{buy}}$, and revenues from sold electricity $P^{\mathrm{sell}}_{t,el}$ at a price of $\mathrm{p}^{\mathrm{sell}}_{el}$. The costs for maintenance CM_u depend on a fraction of the capital expenditure for each energy conversion unit (Equation (A.14)).

We also rewrite the generic energy balance from Equation (3.9) for the energy forms heat, cold, and electricity (el):

$$E^{\mathrm{supply}}_{t,heat} = \sum_{u\in U}\sum_{g\in G_u} V_{t,u,heat,g} - \sum_{ac\in AC} U_{t,ac} \quad \forall t \in T \tag{A.11}$$

$$E^{\mathrm{supply}}_{t,cold} = \sum_{u\in U}\sum_{g\in G_u} V_{t,u,cold,g} \quad \forall t \in T \tag{A.12}$$

$$E^{\mathrm{supply}}_{t,el} = \sum_{chp\in CHP}\sum_{g\in G_{chp}} V_{t,chp,el,g} + P^{\mathrm{buy}}_{t,el} - P^{\mathrm{sell}}_{t,el} - \sum_{cc\in CC} U_{t,cc} \quad \forall t \in T. \tag{A.13}$$

The index g denotes the linear section in which the energy conversion unit is operated. The supplied heat $V_{t,u,heat,g}$ is decreased by the heat demand of the absorption chillers $U_{t,ac}$ (Equation (A.11)). The supplied electricity $V_{t,chp,el,g}$ is decreased by the electricity that is sold to the grid $P^{\mathrm{sell}}_{t,el}$ and consumed by the compression chillers $U_{t,cc}$, and increased by the electricity that is purchased from the grid $P^{\mathrm{buy}}_{t,el}$ (Equation (A.13)).

The considered maintenance costs CM_u depend on the investment costs of each energy conversion unit u with fixed $\mathrm{C}^{\mathrm{fix}}_{u,h}$ and variable $\mathrm{C}^{\mathrm{var}}_{u,h}$ cost:

$$CM_u = \mathrm{mf}_u \cdot \sum_{h\in H_u} \left(\kappa_{u,h} \cdot \mathrm{C}^{\mathrm{fix}}_{u,h} + V^{\mathrm{N}}_{u,h} \cdot \mathrm{C}^{\mathrm{var}}_{u,h}\right) \quad \forall u \in U. \tag{A.14}$$

To model economy of scale, the capital expenditure of each energy conversion unit is piecewise linearized. $\kappa_{u,h}$ equals 1 for the linear segment h in which the energy conversion unit u is installed. The segments for the linearization h of the energy conversion units thermal capacity $V^{\mathrm{N}}_{u,h}$ are bound by a lower and upper bound $\mathrm{capacity}^{\mathrm{th}}_{u,h}$ for the corresponding segment h:

$$V^{\mathrm{N}}_{u,h} \geq \kappa_{u,h} \cdot \mathrm{capacity}^{\mathrm{th}}_{u,h} \quad \forall u \in U, h \in H_u \tag{A.15}$$

$$V^{\mathrm{N}}_{u,h} \leq \kappa_{u,h} \cdot \mathrm{capacity}^{\mathrm{th}}_{u,h+1} \quad \forall u \in U, h \in H_u. \tag{A.16}$$

Each energy conversion unit is installed ($y_u = 1$) with a fixed thermal capacity. The installation is ensured by the following constraint:

$$\sum_{h\in H_u} \kappa_{u,h} = y_u \quad \forall u \in U. \tag{A.17}$$

When an energy conversion unit is operated in one time step t, the energy conversion unit needs to be installed. The binary variable $o_{t,u,g}$ indicates in which piecewise-linearized operation range an energy conversion unit runs in time step t. An energy conversion unit can only run in one operation range in each time step t:

$$\sum_{g \in G_u} o_{t,u,g} \leq y_u \quad \forall t \in T, u \in U. \tag{A.18}$$

The supplied energy from energy conversion unit $V_{t,u,e,g}$ is limited by the capacity and the minimal and maximal part load. These part-load limits are calculated by multiplying the time-dependent capacity $\xi_{t,u,g}$ (Equation (A.24)-(A.27)) with the minimal and maximal relative thermal output energy $\mathrm{v}_{u,g}^{\min}$ and $\mathrm{v}_{u,g}^{\max}$. $\xi_{t,u,g}$ is 0 if the energy conversion unit is off and equals the capacity if operated. For the electrical output of CHP engines, the thermal capacity needs to be converted into the electrical capacity:

$$\mathrm{v}_{u,g}^{\min} \cdot \xi_{t,u,g} \leq V_{t,u,e,g} \leq \mathrm{v}_{u,g}^{\max} \cdot \xi_{t,u,g} \quad \forall u \in U, t \in T, g \in G, e \in E \tag{A.19}$$

$$V_{t,chp,el,g} \leq \mathrm{v}_{chp,el,g}^{\max} \cdot \xi_{t,chp,g} \cdot \frac{\eta_{chp} - \eta_{chp}^{th.}}{\eta_{chp}^{th.}} \quad \forall chp \in CHP, t \in T, g \in G \tag{A.20}$$

$$V_{t,chp,el,g} \geq \mathrm{v}_{chp,el,g}^{\min} \cdot \xi_{t,chp,g} \cdot \frac{\eta_{chp} - \eta_{chp}^{th.}}{\eta_{chp}^{th.}} \quad \forall chp \in CHP, t \in T, g \in G. \tag{A.21}$$

To model part-load performance, the operational curves of input energy $U_{t,u}$ and output energy $V_{t,u,e,g}$ are piecewise linearized:

$$U_{t,u} = \sum_{g \in G_u} \left[\mathrm{u}_{u,g} \cdot \frac{\xi_{t,u,g}}{\eta_u^{th.}} + (V_{t,u,e,g} - \xi_{t,u,g} \cdot \mathrm{v}_{u,g}) \cdot \left(\frac{du}{dv}\right)_{u,g} \right] \quad \forall t \in T, u \in U. \tag{A.22}$$

For the CHP engines, an additional constraint is necessary for electrical power part-load modeling:

$$U_{t,chp} = \sum_{g \in G_{chp}} \left(\mathrm{u}_{chp,g}^{\mathrm{el}} \cdot \frac{\xi_{t,chp,g}}{\eta_{chp}^{th.}} + \left[V_{t,chp,el.,g} - \xi_{t,chp,g} \cdot \mathrm{v}_{chp,g}^{el} \cdot \left(\frac{\eta_{chp} - \eta_{chp}^{th.}}{\eta_{chp}^{th.}} \right) \right] \cdot \left(\frac{\mathrm{du}}{\mathrm{dv}} \right)_{chp,g}^{el} \right) \quad \forall chp \in CHP, t \in T. \tag{A.23}$$

We apply the method proposed by Glover (1975) to represent the product of the binary variable $o_{t,u,g}$ and the continuous variable $V_{u,h}^{\mathrm{N}}$, by a the variable $\xi_{t,u,g}$:

$$\xi_{t,u,g} \leq o_{t,u,g} \cdot \max_h \left(\mathrm{capacity}_{u,h}^{\mathrm{th}} \right) \quad \forall u \in U, t \in T, g \in G \tag{A.24}$$

$$\xi_{t,u,g} \geq o_{t,u,g} \cdot \min_{h} \left(\text{capacity}^{\text{th}}_{u,h}\right) \quad \forall u \in U, t \in T, g \in G \tag{A.25}$$

$$\xi_{t,u,g} \leq \sum_{h=h_1}^{h_{\max}-1} V^{\text{N}}_{u,h} \quad \forall u \in U, t \in T, g \in G \tag{A.26}$$

$$\xi_{t,u,g} \geq \sum_{h=h_1}^{h_{\max}-1} V^{\text{N}}_{u,h} + \left(o_{t,u,g} - 1\right) \cdot \max_{h} \left(\text{capacity}^{\text{th}}_{u,h}\right) \quad \forall u \in U, t \in T, g \in G. \tag{A.27}$$

A.2 Equations for changes of the energy demand

In the method presented in Section 5.1, in each repetition of the Stackelberg game, the production system updates the energy demand $E^{\text{new}}_{t,e}$ to minimize the total costs (Equation (5.16)). The changes of the energy demand $\Delta E^{\text{upper}}_{t,e}$, $\Delta E^{\text{curr+}}_{t,e}$, $\Delta E^{\text{curr-}}_{t,e}$, and $\Delta E^{\text{lower}}_{t,e}$ describe the changes of the updated energy demand $E^{\text{new}}_{t,e}$ compared to the current energy demand $E^{curr}_{t,e}$ within each energy price band (Figure 5.2). For each price band, the change of the energy demand is defined separately. The definition for the energy price band for large decrease is:

$$\Delta E^{lower}_{t,e} = (E^{\text{new}}_{t,e} - E^{\text{lb1}}_{t,e}) \cdot \alpha_{t,e,1} \quad \forall t \in T, e \in E. \tag{A.28}$$

The definition for the energy price band for small decrease is:

$$\Delta E^{curr-}_{t,e} = (E^{\text{new}}_{t,e} - E^{curr}_{t,e}) \cdot \alpha_{t,e,2} + (E^{\text{lb1}}_{t,e} - E^{curr}_{t,e}) \cdot \alpha_{t,e,1} \quad \forall t \in T, e \in E. \tag{A.29}$$

The definition for the energy price band for small increase is:

$$\Delta E^{curr+}_{t,e} = (E^{\text{new}}_{t,e} - E^{curr}_{t,e}) \cdot \alpha_{t,e,3} + (E^{\text{ub1}}_{t,e} - E^{curr}_{t,e}) \cdot \alpha_{t,e,4} \quad \forall t \in T, e \in E. \tag{A.30}$$

The definition for the upper energy price band is:

$$\Delta E^{\text{upper}}_{t,e} = (E^{\text{new}}_{t,e} - E^{\text{ub1}}_{t,e}) \cdot \alpha_{t,e,4} \quad \forall t \in T, e \in E. \tag{A.31}$$

In Equation (A.28)-(A.31), the binary variables $\alpha_{t,e,k}$ indicate in which energy price band the energy demand is changed. Thus, for every time step it holds:

$$\sum_{k=1}^{4} \alpha_{t,e,k} = 1 \quad \forall t \in T, e \in E. \tag{A.32}$$

In our MILP model, Equation (A.28)-(A.31) are linearized by the method proposed by Glover (1975).

Appendix B

Parameters of the case studies

In this chapter, we present the parameters used in the case studies for the production systems.

Major parts of this chapter are reproduced by permission of Elsevier with modifications from the complete Elsevier from:

> Leenders, L., Bahl, B., Hennen, M., and Bardow, A. (2019a). Coordinating scheduling of production and utility system using a Stackelberg game. *Energy*, 175, 1283-1295.
>
> Leenders, L., Bahl, B., Lampe, M., Hennen, M., and Bardow, A. (2019b). Optimal design of integrated batch production and utility systems. *Computers & Chemical Engineering*, 128, 496-511.
>
> The author of this thesis contributed to the development and implementation of the methods as well as the calculation and interpretation of the results, wrote the draft of the papers and is their principal author.

B.1 Parameters of the case study I in Section 3.2.1

In this Section, we present the value of the parameters used in the case study in Section 3.2.1. The case study is based on Kondili et al. (1993). The parameter values of the energy system are equal to Voll et al. (2013). The other parameter values are given in Table B.1, Table B.2 and Table B.3.

Table B.1: Parameter values of case study I depending on the equipment.

parameter	E01	E02	E03	E04
C_j^{fix} /€	100000	150000	120000	150000
C_j^{var} /(€/t)	200	500	0	300
$OC_{i,j}^{\text{fix}}$ /(€/h)	60	60	60	60
$OC_{i,j}^{\text{var}}$/(€/(t·h))	2	2	2	2
V_j^{ub} /t	80	90	70	100
V_j^{lb} /t	10	30	70	40
$V_{s,j}^{ub}$ /t	300	300	300	300
$V_{s,j}^{lb}$ /t	0	0	0	0
$C_{s,j}^{\text{piping, fix}}$ /€	10000	10000	10000	10000
$C_{s,j}^{\text{piping, var}}$/(€/t)	500	500	500	500

Table B.2: Parameter values of the case study I depending on the storage.

parameter	S01	S02	S03	S05	S06	S07	S08	S09	S10
C_s^{fix} /€	10000	10000	10000	30000	15000	10000	20000	5000	10000
C_s^{var} /(€/t)	100	100	100	100	100	100	200	100	200
V_s^{ub} /t	10000	10000	10000	300	600	700	1000	10000	10000
V_s^{lb} /t	0	0	0	10	10	50	10	10	10

Table B.3: Equipment, scheduling data and energy demands of the production system for case study I. E: equipment; T: task; El: electricity demand; τ_1: first time step of task; τ_2: second time step of task.

equipment (available capacity /t)	suitable tasks (duration/h)	energy demand	fixed energy demand /kWh		variable energy demand /(kWh/t)	
			τ_1	τ_2	τ_1	τ_2
E1 (10-80)	T1(1)	Heat	120		30	
E2 (30-90)	T2(2)	Cold	60	80	15	20
		El	40	40	10	10
	T3(2)	Heat	80	60	20	15
		El	20	20	5	5
	T4(1)	Cold	80		20	
		El	20		5	
E3 (70)	T2(2)	Cold	60	80	15	20
		El	40	40	10	10
	T3(2)	Heat	80	60	20	15
		El	20	20	5	5
	T4(1)	Cold	80		20	
		El	20		5	
E4 (40-100)	T5(2)	Heat	120	160	30	40
		El	28	12	7	3

B.2 Parameters of the case study II in Section 3.2.2

In this section, we present the parameter values used in case study II in Section 3.2.2. The case study is based on Kallrath (2002). The parameter values of the energy system are equal to Voll et al. (2013). The other parameter values are given in Table B.4, Table B.5, and Table B.6.

Table B.4: Parameter values of case study II depending on the equipment.

parameter	E01	E02	E03	E04	E05	E06	E07	E08	E09
C_j^{fix} / $10^3 \cdot$ €	30	30	30	30	30	30	30	30	30
C_j^{var} /(€/t)	500	500	500	500	500	500	500	500	500
$OC_{i,j}^{\text{fix}}$ /(€/h)	60	60	60	60	60	60	60	60	60
$OC_{i,j}^{\text{var}}$/(€/(t·h))	2	2	2	2	2	2	2	2	2
V_j^{ub} /t	100	100	500	300	300	300	300	300	300
V_j^{lb} /t	0	0	0	0	0	0	0	0	0
$V_{s,j}^{ub}$ /t	300	300	300	300	300	300	300	300	300
$V_{s,j}^{lb}$ /t	0	0	0	0	0	0	0	0	0
$C_{s,j}^{\text{piping, fix}}$ /€	1	1	1	1	1	1	1	1	1
$C_{s,j}^{\text{piping, var}}$/(€/t)	10	10	10	10	10	10	10	10	10

Table B.5: Parameter values of case study II depending on the storage.

parameter	S00	S01	S02	S03	S04	S05	S06	S07	S08	S09	S10	S11	S12	S13	S14	S15	S16	S17	S18
C_s^{fix} /€	100	100	100	100	100	100	100	100	100	100	100	100	100	100	100	100	100	100	100
C_s^{var} /(€/t)	10	10	10	10	10	100	100	100	100	100	100	100	100	100	100	100	100	100	100
V_s^{ub} /t	1000	1000	1000	1000	1000	500	500	500	500	500	500	500	500	500	500	500	500	500	500
V_s^{lb} /t	0	0	0	0	0	0	0	0	0	0	0	0	0	0	0	0	0	0	0

Table B.6: Equipment, scheduling data and energy demands of the production system for case study II. E: equipment; T: task; El: electricity demand; τ_i: ith time step of a task.

equipment (available capacity /t)	suitable task (duration/h)	energy demand	fixed energy demand /kWh						variable energy demand /(kWh/t)					
			τ_1	τ_2	τ_3	τ_4	τ_5	τ_6	τ_1	τ_2	τ_3	τ_4	τ_5	τ_6
E01	T01(2)	Cold	120	120					30	30				
		El	60	60					15	15				
E02	T02(4)	Heat	40	40	60	40			10	10	15	10		
E03	T03(2)	Heat	20	40					5	10				
E04	T04(4)	Heat	160	140	178	170			40	35	44.5	42.5		
		El	20	40	20	100			5	10	5	25		
	T05(4)	Heat	30	20	30	28			7.5	5	7.5	7		
		El	0	40	0	20			0	10	0	5		
	T06(4)	Heat	200	240	198	230			50	60	49.5	57.5		
		El	40	60	40	20			10	15	10	5		
	T07(4)	Heat	220	180	258	250			55	45	64.5	62.5		
		El	40	40	40	20			10	10	10	5		
E05	T08(6)	Cold	120	120	98	118	120	120	30	30	24.5	29.5	30	30
		El	20	20	20	20	40	20	5	5	5	5	10	5
	T09(6)	Cold	138	158	180	190	180	160	34.5	39.5	45	47.5	45	40
		El	40	40	20	40	40	40	10	10	5	10	10	10
E06	T10(4)	Cold	38	40	20	40			9.5	10	5	10		
		El	80	60	60	80			20	15	15	20		
	T11(5)	Cold	38	38	20	20	40		9.5	9.5	5	5	10	
		El	80	0	20	20	40		20	0	5	5	10	
	T12(6)	Cold	138	98	160	100	130	170	34.5	24.5	40	25	32.5	42.5
		El	40	40	20	40	40	40	10	10	5	10	10	10
E07	T10(5)	Cold	58	20	20	40	40		14.5	5	5	10	10	
		El	40	60	20	40	40		10	15	5	10	10	
	T11(6)	Cold	38	18	20	20	20	20	9.5	4.5	5	5	5	5
		El	80	0	0	20	20	0	20	0	0	5	5	
	T12(6)	Cold	138	98	160	160	100	140	34.5	24.5	40	40	25	35
		El	40	40	60	40	20	20	10	10	15	10	5	5
E08	T13(6)	Heat	20	40	42	20	0	0	5	10	10.5	5	0	0
		El	20	60	40	100	0	0	5	15	10	25	0	0
	T14(4)	Heat	240	204	238	230			60	51	59.5	57.5		
		El	40	0	20	40			10	0	5	10		
	T15(4)	Heat	40	24	18	30			10	6	4.5	7.5		
		El	60	80	40	60			15	20	10	15		
	T16(6)	Heat	40	40	62	40	20	60	10	10	15.5	10	5	15
		El	0	40	40	160	100	100	0	10	10	40	25	25
	T17(6)	Heat	200	220	162	180	200	120	50	55	40.5	45	50	30
		El	0	20	20	40	20	20	0	5	5	10	5	5
E09	T13(6)	El	40	60	62	60	80	80	10	15	15.5	15	20	20
	T14(6)	El	200	240	138	230	190	180	50	51	34.5	57.5	47.5	45
	T16(6)	El	100	100	122	100	100	120	25	25	30.5	25	25	30
	T17(6)	El	240	160	202	220	200	180	60	40	50.5	55	50	45

B.3 Parameters of the case study I in Section 5.2.1

In Table B.7, Table B.8, and Table B.9, the parameters for the production system in case study I are presented.

Table B.7: Energy demand of the tasks for case study I. E: Equipment; T: Task; El: Electricity demand; τ_i: ith time step of a task.

equipment	task	energy demand	fixed energy demand /kWh						variable energy demand /kWh/t					
			τ_1	τ_2	τ_3	τ_4	τ_5	τ_6	τ_1	τ_2	τ_3	τ_4	τ_5	τ_6
E01	T01	El	300	300					75	75				
E02	T02	Heat	200	200	300	200			50	50	75	50		
E03	T03	Heat	100	200					25	50				
E04	T04	Heat	800	700	890	850			200	175	222.5	212.5		
		El	100	200	100	500			25	50	25	125		
	T05	Heat	150	100	150	140			37.5	25	37.5	35		
		El	0	200	0	100			0	50	0	25		
	T06	Heat	1000	1200	990	1150			250	300	247.5	287.5		
		El	200	300	200	100			50	75	50	25		
	T07	Heat	1100	900	1290	1250			275	225	322.5	312.5		
		El	200	200	200	100			50	50	50	25		
E05	T08	El	100	100	100	100	200	100	25	25	25	25	50	25
	T09	El	200	200	100	200	200	200	50	50	25	50	50	50
E06	T10	El	400	300	300	400			100	75	75	100		
	T11	El	400	0	100	100	200		100	0	25	25	50	
	T12	El	200	200	100	200	200	200	30	50	25	50	50	50
E07	T10	El	200	300	100	200	200		50	75	25	50	50	
	T11	El	400	0	0	100	100		100	0	0	25	25	
	T12	El	200	200	300	200	100	100	50	50	75	50	25	25
E08	T13	Heat	100	200	210	100			25	50	52.5	25		
		El	100	300	200	500			25	75	50	125		
	T14	Heat	1200	1020	1190	1150			300	255	297.5	287.5		
		El	200	0	100	200			50	0	25	50		
	T15	Heat	200	120	90	150			50	30	22.5	37.5		
		El	300	400	200	300			75	100	50	75		
	T16	Heat	200	200	310	200	100	300	50	50	77.5	50	25	75
		El	0	200	200	800	500	500	0	50	50	200	125	125
	T17	Heat	1000	1100	810	900	1000	600	250	275	202.5	225	250	150
		El	0	100	100	200	100	100	0	25	25	50	25	25
E09	T13	El	200	300	310	300	400	400	50	75	77.5	75	100	100
	T14	El	1000	1020	690	1150	950	900	250	255	172.5	287.5	237.5	225
	T16	El	500	500	610	500	500	600	125	125	152.5	125	125	150
	T17	El	1200	800	1010	1100	1000	900	200	300	252.5	275	250	225

Table B.8: Operational costs of the equipment for case study I.

parameter	E01	E02	E03	E04	E05	E06	E07	E08	E09
$OC_{i,j}^{fix}$ /(€/h)	60	60	60	60	60	60	60	60	60
$OC_{i,j}^{var}$/(€/(t·h))	2	2	2	2	2	2	2	2	2

Table B.9: Storage size for case study I.

parameter	V_s /t
S00	500
S01	30
S02	30
S03	15
S04	30
S05	0
S06	10
S07	10
S08	10
S09	0
S10	0
S11	10
S12	0
S13	10
S14	500
S15	400
S16	500
S17	500
S18	500

B.4 Parameters of the case study II in Section 5.2.2

In Table B.10, Table B.11, and Table B.12, the parameters for the production system in case study II are presented.

Table B.10: Energy demand of the tasks for case study II. E: equipment; T: task; El: electricity demand; τ_i: ith time step of task.

equipment	task	energy demand	fixed energy demand /kWh		variable energy demand /kWh/t	
			τ_1	τ_2	τ_1	τ_2
E01	T01	Heat	200		50	
E02	T02	Heat	40	200	10	50
		El	80	20	20	5
	T03	Heat	80	60	20	15
		El	60	100	15	25
	T04	Heat	160		40	
		El	60		15	
E03	T02	Heat	120	160	30	40
		El	80	40	20	10
	T03	Heat	80	60	20	15
		El	20	80	5	20
	T04	Heat	200		50	
		El	20		5	
E04	T05	Heat	120	160	30	40
		El	60	20	15	5

Table B.11: Operational costs of the equipment for case study II.

parameter	E01	E02	E03	E04
$OC^{fix}_{i,j}$ /(€/h)	60	60	60	60
$OC^{var}_{i,j}$ /(€/(t·h))	2	2	2	2

Table B.12: Storage size for case study II.

parameter	S01	S02	S03	S04	S05	S06	S07	S08	S09	S10
V_s /t	200	200	200	200	200	200	20,000	200	200	20,000

Appendix C

Additional results for the case studies in Section 5.2

In this chapter, we present additional results for the case studies in Section 5.2. We present the energy demand of case study II as well as results for a variation of the initial energy demand.

Major parts of this chapter are reproduced by permission of Elsevier with modifications from the complete Elsevier source from:

Leenders, L., Bahl, B., Hennen, M., and Bardow, A. (2019a). Coordinating scheduling of production and utility system using a Stackelberg game. *Energy*, 175, 1283-1295.

The author of this thesis contributed to the development and implementation of the method as well as the calculation and interpretation of the results, wrote the draft of the paper and is its principal author.

C.1 Energy demand of case study II

In Figure C.1, the energy demand of case study II is presented. The energy demand profiles differ in the approaches. The different energy demand profile results in different optimal operations of the energy system and thereby lead to cost savings in the energy system. In case study II, 87.9 % of the electricity is supplied by the CHP engines in the sequential approach, whereas in the repeated Stackelberg game 97.5 %, of the electricity is supplied by the CHP engines.

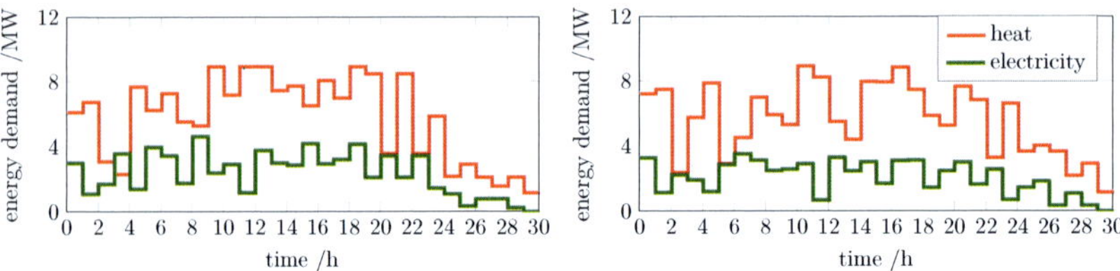

Figure C.1: Energy demand of the sequential approach (left) and the proposed method (right) for case study II.

C.2 Variation of initial energy prices

In Section 5.2, initial energy prices were set to zero for both case studies. These initial energy prices correspond to the case that no information on the energy prices is available. Still, the results show that the proposed Stackelberg game leads to significant cost savings.

To show the influence of different initial energy prices on the optimal schedules, we investigate three sets of initial energy prices. These initial energy prices are assumed to be constant over time. The three energy prices scenarios are:

- near-optimal scenario
  - mean value over time of the energy prices obtained from the optimal solution of the Stackelberg game that started with energy prices equal zero. These prices represent a very good estimate of constant energy prices in practice
- high heat price scenario
  - initial energy prices with a heating price equal to the gas price 0.06 €/kWh and low electricity price 0.04 €/kWh
- high electricity price scenario

– initial energy prices with a low heating price 0.02 €/kWh and electricity price equal to half of the electricity price from the grid 0.08 €/kWh.

C.2.1 Case study I

For all investigated initial energy prices, the method reduces the total cost between 7 % and 9.1 % (Figure C.2). Thus, the solutions depend on the initial energy prices, but in all cases, the proposed method decreases total cost significantly even for the very different initial energy prices.

For the near-optimal scenario, the first run of the Stackelberg game yields lower total cost than for zero initial energy prices (Figure C.2). This result was expected since the prices were known to be near-optimal. In the two following repetitions of the Stackelberg game, total cost decreases, exploiting that energy prices are no longer constant but defined for each time step. After the second repetition, the total cost increases, and the proposed method aborts.

For the high heat and the high electricity scenarios, the first run of the Stackelberg game yields higher and lower total costs than for zero initial energy prices, respectively. Ultimately, solutions with very similar costs are found.

Thus, the initial energy prices affect the results of the first run of the Stackelberg game significantly because production schedules are chosen that do not approximate the actual cost well. However, in further repetitions, the total cost always decreases compared to the sequential optimization.

In this case study, zero initial energy prices lead to the solution with the lowest total cost, but this result cannot be generalized as shown in the following section.

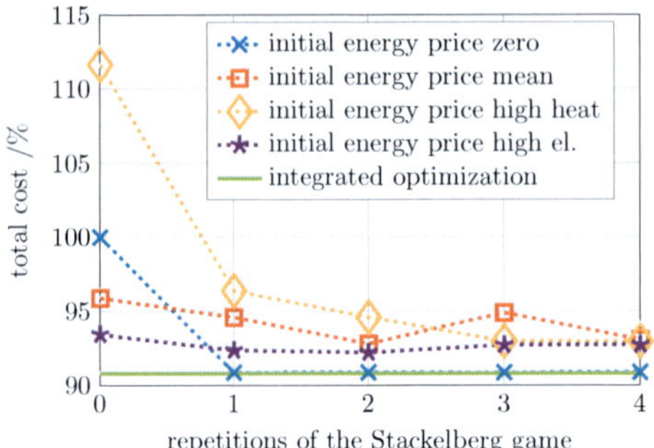

Figure C.2: Total costs compared to the sequential approach (=100 %) for the proposed repeated Stackelberg game with the demand-dependent energy cost for case study I. Here, the initial energy prices are varied. Further repetitions after the abortion of the Stackelberg game are shown.

C.2.2 Case study II

For all investigated initial energy prices, the method reduces the total cost between 4.7 % and 5 % (Figure C.3). Thus, the solutions depend on the initial energy prices, but for all initial energy prices, total cost decreases significantly.

For the near-optimal scenario, the first run of the Stackelberg game results in lower total cost compared to zero initial energy prices. Again, this result was expected since the prices are known to be near-optimal. In the first repetition of the Stackelberg game, the total cost decreases further since the energy prices are no longer constant over time. For the high heat and high electricity price scenario, the first run of the Stackelberg game results in lower total cost compared to zero initial energy prices. In this case study, all initial energy prices result in nearly the same total cost in the first repetition of the Stackelberg game. The total cost is not significantly decreased in the following repetitions.

Concluding both case studies, the initial energy prices significantly affect the total cost of the first run of the Stackelberg game. However, the proposed method reduces the total cost significantly for all initial energy prices.

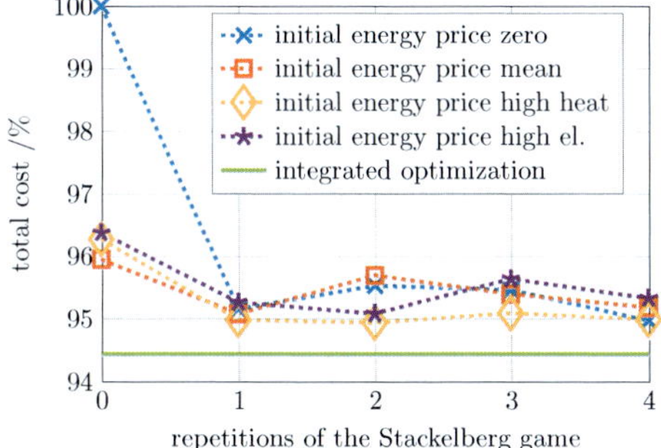

Figure C.3: Total costs compared to the sequential approach (=100 %) for the proposed repeated Stackelberg game with the demand-dependent energy cost for case study II. Here, the initial energy prices are varied. Further repetitions after the abortion of the Stackelberg game are shown.

Appendix D

Additional constraints of the discretization points in the lower-bounding problem

In this section, we provide the additional constraints of discretization points in the lower-bounding problem. By these additional constraints, we identify if the discretization point is valid, i.e., if the free energy conversion units are operated within their operational bounds.

Leenders, L., Hagedorn, D. F., Djelassi, H., Bardow, A., and Mitsos, A. (2022). Bilevel optimization for joint scheduling of production and energy systems. *Optimization & Engineering.*

The author of this thesis contributed to the development and implementation of the method as well as the calculation and interpretation of the results, wrote the draft of the paper and is its principal author.

For every discretization point k and time step t, the heat demand $E^{\text{demand}}_{t,e=\text{heat}}$ of the production system needs to be covered by the heat supply $V^{\text{D}}_{k,t,u,e=\text{heat}}$ of the energy conversion units:

$$\sum_{u \in U} V^{\text{D}}_{k,t,u,e=\text{heat}} = E^{\text{demand}}_{t,e=\text{heat}}, \forall k \in K, t \in T. \tag{D.1}$$

The electricity demand $E^{\text{demand}}_{t,e=\text{el}}$ needs to be covered by the electrical supply $V^{\text{D}}_{k,t,u,e=\text{el}}$ of the energy conversion units:

$$\sum_{u \in U} V^{\text{D}}_{k,t,u,e=\text{el}} = E^{\text{demand}}_{t,e=\text{el}}, \forall k \in K, t \in T. \tag{D.2}$$

The discretization points are defined in Section 7.2.2. In our case study, we consider two energy forms heat and electricity. In the energy system, we model boilers, combined-heat-and-power engines, and the electricity grid. Thus, we define operational limits for energy conversion units supplying heat and electricity. The operational limits for boilers and combined-heat-and-power engines are defined for the energy form heat. The heat and electricity supply of a combined-heat-and-power engine is connected. Thus, the operational limits of the electricity supply can be defined by the operational limit of the heat supply. To identify if the operational limits of the free energy conversion unit hold, big-M formulations are used in the lower-bounding problem. For heat-supplying energy conversion units, the lower operational limit is identified by:

$$\begin{aligned} &V^{\text{D}}_{k,t,u=d^{\text{free}}_{k,f},e=\text{heat}} \leq \text{V}^{\text{min}}_{u=d^{\text{free}}_{k,f},e=\text{heat}} + (1 - \beta^{\text{lower}}_{k,t,f}) \cdot \text{M}^{\text{lower}}_{k,t,f}, \\ &\forall k \in \{K | d^{\text{free}}_{k,f} \in D^{\text{CHP}} \cup D^{Boiler}\}, t \in T, f \in \{1,2\}. \end{aligned} \tag{D.3}$$

The binary variables $\beta^{\text{lower}}_{k,t,f}$ identify if the supplied energy of the free energy conversion unit $V^{\text{D}}_{k,t,u=d^{\text{free}}_{k,f},e}$ is lower than the minimal load $\text{V}^{\text{min}}_{u=d^{\text{free}}_{k,f},e}$. The big-M parameter $M^{\text{lower}}_{k,t,f}$ needs to be chosen such that the heat balance can be fulfilled for every heat demand that the energy system can be supply.

$M^{\text{lower}}_{k,t,f}$ is defined as the maximum heat $\text{M}^{\text{max}}_{e=\text{heat}}$ that can be supplied by the energy system reduced by the heat already supplied by the other energy conversion units $\text{V}^{\text{D}}_{k,t,u=d,e=\text{heat}}$ and the minimal load of the free energy conversion unit $\text{V}^{\text{min}}_{u=d^{\text{free}}_{k,f},e=\text{heat}}$:

$$\begin{aligned} \text{M}^{\text{lower}}_{k,t,f} =& \text{M}^{\text{max}}_{e=\text{heat}} - \sum_{d \in D^{\text{min}}_k \cup D^{\text{max}}_k} \text{V}^{\text{D}}_{k,t,u=d,e=\text{heat}} - \text{V}^{\text{min}}_{u=d^{\text{free}}_{k,f},e=\text{heat}}, \\ &\forall k \in \{K | d^{\text{free}}_{k,f} \in D^{\text{CHP}} \cup D^{Boiler}\}, t \in T, f \in \{1,2\}. \end{aligned} \tag{D.4}$$

To identify if the upper operational limit holds for heat-supplying energy conversion units, the following equation is stated:

$$\begin{aligned} &V^{\mathrm{D}}_{k,t,u=d^{\mathrm{free}}_{k,f},e=\mathrm{heat}} \geq \mathrm{V}^{\max}_{u=d^{\mathrm{free}}_{k,f},e=\mathrm{heat}} - (1-\beta^{\mathrm{upper}}_{k,t,f}) \cdot \mathrm{M}^{\mathrm{upper}}_{k,t,f}, \\ &\forall k \in \{K|d^{\mathrm{free}}_{k,f} \in D^{\mathrm{CHP}} \cup D^{Boiler}\}, t \in T, f \in \{1,2\}. \end{aligned} \tag{D.5}$$

The binary variable $\beta^{\mathrm{upper}}_{k,t,f}$ identifies if the supplied energy of the free energy conversion unit $V^{\mathrm{D}}_{k,t,u=d^{\mathrm{free}}_{k,f},e}$ is higher than the maximal load $\mathrm{V}^{\max}_{u=d^{\mathrm{free}}_{k,f},e}$. The big-M parameter $\mathrm{M}^{\mathrm{upper}}_{k,t,f}$ needs to be chosen such that the heat balance can be fulfilled for every heat demand that can be supplied by the energy system. Therefore, $\mathrm{M}^{\mathrm{upper}}_{k,t,f}$ is defined as the minimal heat that can be supplied by the energy system increased by the heat already supplied by the other energy conversion units, the maximal heat load of the largest combined-heat-and-power engine $\mathrm{V}^{\max}_{e=\mathrm{heat}}$, and the maximal load of the free energy conversion unit $\mathrm{V}^{\max}_{u=d^{\mathrm{free}}_{k,f},e=\mathrm{heat}}$:

$$\begin{aligned} \mathrm{M}^{\mathrm{upper}}_{k,t,f} =& -\mathrm{M}^{\min}_{\mathrm{heat}} + \sum_{d \in D^{\min}_k \cup D^{\max}_k} \mathrm{V}^{\mathrm{D}}_{k,t,u=d,e=\mathrm{heat}} + \mathrm{V}^{\max}_{e=\mathrm{heat}} \\ &+ \mathrm{V}^{\max}_{u=d^{\mathrm{free}}_{k,f},e=\mathrm{heat}}, \forall k \in \{K|d^{\mathrm{free}}_{k,f} \in D^{\mathrm{CHP}} \cup D^{Boiler}\}, t \in T, f \in \{1,2\}. \end{aligned} \tag{D.6}$$

The parameter $\mathrm{M}^{\min}_{\mathrm{heat}}$ is the minimal heat that can be supplied and is commonly 0.

The maximal heat load of the largest combined-heat-and-power engine $\mathrm{V}^{\max}_{e=\mathrm{heat}}$ is required for the case a combined-heat-and-power engine unit is selected as free energy conversion unit for electricity. Thus, with the parameter $\mathrm{M}^{\mathrm{upper}}_{k,t,f}$, the provided heat of the free energy conversion unit can be reduced such that the minimal heat supply can be reached in the heat balance.

In our model, we consider two types of electricity grids: electricity purchase and electricity sale. If an electricity grid is the free energy conversion unit, we also need to check if the operational limits hold. The lower operational limit of the electricity grids is 0. Whether the lower operational limit holds is identified by:

$$\begin{aligned} V^{\mathrm{D}}_{k,t,u=d^{\mathrm{free}}_{k,f},e=\mathrm{el}} \leq & (1-\beta^{\mathrm{lower}}_{k,t,f}) \cdot \mathrm{M}^{\max}_{e=\mathrm{el}}, \\ & \forall k \in \{K|d^{\mathrm{free}}_{k,f} \in D^{\mathrm{grid}}\}, t \in T, f \in \{1,2\}. \end{aligned} \tag{D.7}$$

The big-M parameter $\mathrm{M}^{\max}_{e=\mathrm{el}}$ is the maximum electricity that the production system could demand from the energy system. For the lower operational limit, Equation (D.7) checks if $V^{\mathrm{D}}_{k,t,u=d^{\mathrm{free}}_{k,f},e=\mathrm{el}}$ is non-negative.

The upper operational limit is never exceeded for electricity grids:

$$\beta_{k,t,f}^{\text{upper}} = 0, \forall k \in \{K | d_{k,f}^{\text{free}} \in D^{\text{grid}}\}, t \in T, f \in \{1,2\}. \tag{D.8}$$

Appendix E

Extended proof for multiple energy forms

In this section, we extend the proof from Section 7.1.2 to multiple energy forms.

Leenders, L., Hagedorn, D. F., Djelassi, H., Bardow, A., and Mitsos, A. (2022). Bilevel optimization for joint scheduling of production and energy systems. *Optimization & Engineering*.

The author of this thesis contributed to the development and implementation of the method as well as the calculation and interpretation of the results, wrote the draft of the paper and is its principal author.

Proposition 2. *Let the lower-level problem be given for multiple energy forms e and one time step t as*

$$\begin{aligned}
\min_{\boldsymbol{V}_t,\boldsymbol{o}_t} f_t^l(\boldsymbol{V}_t, \boldsymbol{o}_t) &= \boldsymbol{c}_t^{L,ES,V^T} \boldsymbol{V}_t + \boldsymbol{c}_t^{L,ES,o^T} \boldsymbol{o}_t \\
s.t. \sum_{u \in U} V_{t,u,e} &= E_{t,e}^{demand}, \forall e \in E \\
V_{t,u,e}^{-}(o_{t,u}) \leq V_{t,u,e} &\leq V_{t,u,e}^{+}(o_{t,u}), \forall u \in U, e \in E \\
a_u \cdot o_{t,u} &= b_u \cdot V_{t,u,e_1} - V_{t,u,e_2}, \forall (u,e_1,e_2) \in C.
\end{aligned} \tag{E.1}$$

Therein, $\boldsymbol{V}_t$ is the vector of energy supplied by the energy conversion units u in all energy forms $e \in E$, and $\boldsymbol{o}_t$ is the vector of the operational state of each energy conversion unit u.

For this problem, there exists an optimal solution in which the number of energy conversion units not operated at one of their operating limits is less or equal to the number of energy forms e considered. More precisely, let Problem (E.1) be feasible and let $(\boldsymbol{o}_t^*, \boldsymbol{V}_{t,e}^*)$ be globally optimal in Problem (E.1). Then there exists a solution $(\boldsymbol{o}_t^*, \tilde{\boldsymbol{V}}_{t,e})$ that is also globally optimal in Problem (E.1) and for which there exists at most $|E|$ elements $u \in U$ such that

$$V_{t,u,e}^{-}(\boldsymbol{o}_t^*) < \tilde{V}_{t,u,e} < V_{t,u,e}^{+}(\boldsymbol{o}_t^*). \tag{E.2}$$

Note that a connecting energy conversion unit possesses two energy flows that are subject to operational limits. Nevertheless, each connecting energy conversion unit is only counted once for the purposes of this proposition.

Proof. Consider the case that $(\mathbf{o}_t^*, \mathbf{V}_{t,e}^*)$ is given such that for at most $|E|$ elements of $u \in U$, the following equation holds:

$$\mathrm{V}_{t,u,e}^{-}(\mathbf{o}_t^*) < V_{t,u,e}^* < \mathrm{V}_{t,u,e}^{+}(\mathbf{o}_t^*). \tag{E.3}$$

Then, the result is proven immediately with $\tilde{\mathbf{V}}_{t,e} = \mathbf{V}_{t,e}^*$.

In the following, we prove that if we consider two energy forms e_1 and e_2, the maximum number of energy conversion units at their operational limits can be reduced to at maximum two in an optimal solution. This result can then be generalized to an arbitrary number of energy forms e if we iteratively consider two energy forms and reduce the number of the energy conversion units not at their operational limit to at maximum two.

For the case of the two energy forms e_1 and e_2, energy conversion units can either output energy form e_1 or e_2, or if they are a connecting energy conversion unit, they output both energy forms e_1 and e_2. The energy forms e output by an energy conversion unit u is given by the set E_u. Let $u_1, u_2, u_3 \in U, u_1 \neq u_2 \neq u_3$ be given such that

$$\mathrm{V}^{-}_{t,u,e}(\mathbf{o}^*_t) < V^*_{t,u,e} < \mathrm{V}^{+}_{t,u,e}(\mathbf{o}^*_t), \quad \forall u \in \{u_1, u_2, u_3\}, e \in E_u \tag{E.4}$$

and

$$V^*_{t,u,e} = \mathrm{V}^{-}_{t,u,e}(\mathbf{o}^*_t) \vee V^*_{t,u,e} = \mathrm{V}^{+}_{t,u,e}(\mathbf{o}^*_t), \quad \forall u \in U \setminus \{u_1, u_2, u_3\}, e \in E_u. \tag{E.5}$$

To prove Proposition 2, we distinguish 4 cases that can occur:

- Case 1: All three energy conversion units output a single energy form.
- Case 2: One energy conversion unit is a connecting energy conversion unit, and the other two energy conversion units output a single energy form.
- Case 3: Two energy conversion units are a connecting energy conversion unit, and the other energy conversion unit outputs only a single energy form.
- Case 4: All three energy conversion units are connecting energy conversion units.

In the following, we use the fact that if a connecting energy conversion unit reaches the operational limit in one energy form, also the operational limit in the other energy form is reached. Thus, the operational limits of a connecting energy conversion unit u outputting energy forms e_1 and e_2 can be written in terms of the limits in one energy form using Equation (7.10):

$$\begin{aligned} &\mathrm{V}^{-}_{t,u,e_1}(o_{t,u}) \leq V_{t,u,e_1} \leq \mathrm{V}^{+}_{t,u,e_1}(o_{t,u}) \\ &b_u \cdot \mathrm{V}^{-}_{t,u,e_1}(o_{t,u}) - a_u \leq V_{t,u,e_2} \leq b_u \cdot \mathrm{V}^{+}_{t,u,e_1}(o_{t,u}) - a_u \end{aligned} \tag{E.6}$$

Hence, we only have to consider the operational limits in one energy form.

Now, we prove Proposition 2 for the four cases:

Case 1

Since none of the energy conversion units u_1, u_2, and u_3 are a connecting energy conversion unit, we can apply the procedure from the proof of Proposition 1 for the two energy conversion units producing the same energy form e. Thereby, the proof for this case is given.

Case 2

If the two energy conversion units that output a single energy form output the same energy form, we apply the procedure from Proposition 1. As a result, only two energy conversion units are not operated at their operational limits.

Otherwise, energy conversion unit u_1 outputs energy form e_1, energy conversion unit u_2 is a connecting energy conversion unit outputting energy forms e_1 and e_2, and energy conversion unit u_3 outputs only energy form e_2.

From construction of Equation (E.1), it follows that all elements of the set

$$\begin{aligned} M = \{(\mathbf{o}_t^*, \mathbf{V}_{t,e}) | V_{t,u,e} = V_{t,u,e}^*, \forall u \in U \setminus \{u_1, u_2, u_3\} \\ \wedge \mathrm{V}_{t,u,e=e_1}^-(\mathbf{o}_t^*) \leq V_{t,u,e=e_1} \leq \mathrm{V}_{t,u,e=e_1}^+(\mathbf{o}_t^*), \forall u \in \{u_1, u_2\} \\ \wedge \mathrm{V}_{t,u,e=e_2}^-(\mathbf{o}_t^*) \leq V_{t,u,e=e_2} \leq \mathrm{V}_{t,u,e=e_2}^+(\mathbf{o}_t^*), \forall u \in \{u_3\} \\ \wedge \mathrm{a}_u \cdot o_{t,u}^* = \mathrm{b}_u \cdot V_{t,u,e=e_1} - V_{t,u,e=e_2}, \forall u \in \{u_2\} \\ \wedge V_{t,u=u_1,e=e_1} + V_{t,u=u_2,e=e_1} = V_{t,u=u_1,e=e_1}^* + V_{t,u=u_2,e=e_1}^* \\ \wedge b_{u=u_2} \cdot V_{t,u=u_2,e=e_1} + V_{t,u=u_3,e=e_2} = \\ b_{u=u_2} \cdot V_{t,u=u_2,e=e_1}^* + V_{t,u=u_3,e=e_2}^* \} \end{aligned} \tag{E.7}$$

are feasible in Equation (E.1). $\mathbf{V}_{t,e}^*$ is an optimal solution of an linear program on a facet of the feasible set. Accordingly, all points on that facet (points in M) are optimal in Equation (E.1). Note that in the last equation of M the constant part $a_{u=u_2}$ drops because it is written on the left and right side of the equation.

Following the idea of Equation (7.17), we identify the change in the supplied energy until one energy conversion unit reaches its operational limits. For convenience, we define this change in energy form e_1 by:

$$\begin{aligned} \Delta \mathrm{V}_{e=e_1}^{\mathrm{Case2}} = \min\{ & V_{t,u=u_1,e=e_1}^+(\mathbf{o}_t^*) - V_{t,u=u_1,e=e_1}^*, \\ & V_{t,u=u_2,e=e_1}^* - V_{t,u=u_2,e=e_1}^-(\mathbf{o}_t^*), \\ & \frac{1}{b_{u=u_2}} \cdot (V_{t,u=u_3,e=e_2}^+(\mathbf{o}_t^*) - V_{t,u=u_3,e=e_2}^*)\} \end{aligned} \tag{E.8}$$

With $\Delta \mathrm{V}_{e=e_1}^{\mathrm{Case2}}$ we can manipulate the supplied energy by the energy conversion units such that the point $(\mathbf{o}_t^*, \tilde{\mathbf{V}}_{t,e})$ with $\tilde{V}_{t,u,e} = V_{t,u,e}^*, \forall u \in U \setminus \{u_1, u_2, u_3\}, e \in E_u$ and

$$\begin{aligned} \tilde{V}_{t,u=u_1,e=e_1} &= V_{t,u=u_1,e=e_1}^* + \Delta \mathrm{V}_{e=e_1}^{\mathrm{Case2}} \\ \tilde{V}_{t,u=u_2,e=e_1} &= V_{t,u=u_2,e=e_1}^* - \Delta \mathrm{V}_{e=e_1}^{\mathrm{Case2}} \\ \tilde{V}_{t,u=u_2,e=e_2} &= V_{t,u=u_2,e=e_2}^* - b_{u=u_2} \cdot \Delta \mathrm{V}_{e=e_1}^{\mathrm{Case2}} \\ \tilde{V}_{t,u=u_3,e=e_2} &= V_{t,u=u_3,e=e_2}^* + b_{u=u_2} \cdot \Delta \mathrm{V}_{e=e_1}^{\mathrm{Case2}} \end{aligned} \tag{E.9}$$

lies within M and satisfies either $\tilde{V}_{t,u=u_1,e_1} = V^+_{t,u=u_1,e_1}(\mathbf{o}^*_t)$, $\tilde{V}_{t,u=u_2,e_1} = V^-_{t,u=u_2,e_1}(\mathbf{o}^*_t)$ or $\tilde{V}_{t,u=u_3,e_2} = V^+_{t,u=u_3,e_2}(\mathbf{o}^*_t)$, proving the desired property. Then, only two energy conversion units are not operated at their operational limits.

Case 3

Energy conversion unit u_1 outputs energy form e_1, u_2 and u_3 are connecting energy conversion units and output energy forms e_1 and e_2.

From construction of Equation (E.1), it follows that all elements of the set

$$\begin{aligned} M = \{(\mathbf{o}^*_t, \mathbf{V}_{t,e}) | V_{t,u,e} = V^*_{t,u,e}, \forall u \in U \setminus \{u_1, u_2, u_3\} \\ \wedge \mathrm{V}^-_{t,u,e=e_1}(\mathbf{o}^*_t) \leq V_{t,u,e=e_1} \leq \mathrm{V}^+_{t,u,e=e_1}(\mathbf{o}^*_t), \forall u \in \{u_1, u_2, u_3\} \\ \wedge \mathrm{a}_u \cdot o^*_{t,u} = \mathrm{b}_u \cdot V_{t,u,e=e_1} - V_{t,u,e=e_2}, \forall u \in \{u_2, u_3\} \\ \wedge V_{t,u=u_1,e=e_1} + V_{t,u=u_2,e=e_1} + V_{t,u=u_3,e=e_1} = \\ V^*_{t,u=u_1,e=e_1} + V^*_{t,u=u_2,e=e_1} + V^*_{t,u=u_3,e=e_1} \\ \wedge b_{u=u_2} \cdot V_{t,u=u_2,e=e_1} + b_{u=u_3} \cdot V_{t,u=u_3,e=e_1} = \\ b_{u=u_2} \cdot V^*_{t,u=u_2,e=e_1} + b_{u=u_3} \cdot V^*_{t,u=u_3,e=e_1}\} \end{aligned} \tag{E.10}$$

are feasible in Equation (E.1).

$\mathbf{V}^*_{t,e}$ is an optimal solution of a linear program on a facet of the feasible set. Accordingly, all points on that facet (points in M) are optimal in Equation (E.1). Note that again in the last equation of M the constant parts $a_{u=u_2}$ and $a_{u=u_3}$ are eliminated. Following the definition of Equation (E.8), we define the change in energy form e_1 until an energy conversion unit reaches its operational limits:

$$\begin{aligned} \Delta \mathrm{V}^{\mathrm{Case3}}_{e=e_1} = \min\{ & V^+_{t,u=u_1,e=e_1}(\mathbf{o}^*_t) - V^*_{t,u=u_1,e=e_1}, \\ & \frac{b_{u=u_3} - b_{u=u_2}}{b_{u=u_3} \cdot b_{u=u_2}} \cdot (V^*_{t,u=u_2,e=e_2} - V^-_{t,u=u_2,e=e_2}(\mathbf{o}^*_t)), \\ & \frac{b_{u=u_3} - b_{u=u_2}}{b_{u=u_3} \cdot b_{u=u_2}} \cdot (V^+_{t,u=u_3,e=e_2}(\mathbf{o}^*_t) - V^*_{t,u=u_3,e=e_2})\} \end{aligned} \tag{E.11}$$

With $\Delta \mathrm{V}_{e=e_1}^{\text{Case3}}$, we can manipulate the supplied energy by the energy conversion units such that the point $(\mathbf{o}_t^*, \tilde{\mathbf{V}}_{t,e})$ with $\tilde{V}_{t,u,e} = V_{t,u,e}^*, \forall u \in U \setminus \{u_1, u_2, u_3\}, e \in E_u$ and

$$\begin{aligned}
\tilde{V}_{t,u=u_1,e=e_1} &= V_{t,u=u_1,e=e_1}^* + \Delta \mathrm{V}_{e=e_1}^{\text{Case3}} \\
\tilde{V}_{t,u=u_2,e=e_1} &= V_{t,u=u_2,e=e_1}^* - \frac{b_{u=u_3}}{b_{u=u_3} - b_{u=u_2}} \cdot \Delta \mathrm{V}_{e=e_1}^{\text{Case3}} \\
\tilde{V}_{t,u=u_3,e=e_1} &= V_{t,u=u_3,e=e_1}^* + \frac{b_{u=u_2}}{b_{u=u_3} - b_{u=u_2}} \cdot \Delta \mathrm{V}_{e=e_1}^{\text{Case3}} \\
\tilde{V}_{t,u=u_2,e=e_2} &= V_{t,u=u_2,e=e_2}^* - \frac{b_{u=u_2} \cdot b_{u=u_3}}{b_{u=u_3} - b_{u=u_2}} \cdot \Delta \mathrm{V}_{e=e_1}^{\text{Case3}} \\
\tilde{V}_{t,u=u_3,e=e_2} &= V_{t,u=u_3,e=e_2}^* + \frac{b_{u=u_2} \cdot b_{u=u_3}}{b_{u=u_3} - b_{u=u_2}} \cdot \Delta \mathrm{V}_{e=e_1}^{\text{Case3}}
\end{aligned} \tag{E.12}$$

lies within M and satisfies either $\tilde{V}_{t,u=u_1,e_1} = V_{t,u=u_1,e_1}^+(\mathbf{o}_t^*)$, $\tilde{V}_{t,u=u_2,e_2} = V_{t,u=u_2,e_2}^-(\mathbf{o}_t^*)$ or $\tilde{V}_{t,u=u_3,e_2} = V_{t,u=u_3,e_2}^+(\mathbf{o}_t^*)$, proving the desired property. Note that if $b_{u=u_2}{=}b_{u=u_3}$, then u_1 does not change its energy supply.

With these changes in the supplied energy, only two energy conversion units are not operated at their operational limits.

Case 4

This case only occurs if 3 connecting energy conversion units are not at their operational limits and do not have the same conversion factor b_u. If at least 2 of the 3 have the same conversion factor b_u, the special case from Case 3 can be applied, and only the two equal energy conversion units change their energy supply.

Energy conversion units u_1, u_2, and u_3 are connecting energy conversion units, and output energy forms e_1 and e_2. From construction of Equation (E.1), it follows that all elements of the set

$$\begin{aligned}
M = \{(\mathbf{o}_t^*, \mathbf{V}_{t,e}) | V_{t,u,e} = V_{t,u,e}^*, \forall u \in U \setminus \{u_1, u_2, u_3\} \\
&\wedge \mathrm{V}_{t,u,e=e_1}^-(\mathbf{o}_t^*) \leq V_{t,u,e=e_1} \leq \mathrm{V}_{t,u,e=e_1}^+(\mathbf{o}_t^*), \forall u \in \{u_1, u_2, u_3\} \\
&\wedge \mathrm{a}_u \cdot o_{t,u}^* = \mathrm{b}_u \cdot V_{t,u,e=e_1} - V_{t,u,e=e_2}, \forall u \in \{u_1, u_2, u_3\} \\
&\wedge V_{t,u=u_1,e=e_1} + V_{t,u=u_2,e=e_1} + V_{t,u=u_3,e=e_1} = \\
&V_{t,u=u_1,e=e_1}^* + V_{t,u=u_2,e=e_1}^* + V_{t,u=u_3,e=e_1}^* \\
&\wedge b_{u=u_1} \cdot V_{t,u=u_1,e=e_1} + b_{u=u_2} \cdot V_{t,u=u_2,e=e_1} + b_{u=u_3} \cdot V_{t,u=u_3,e=e_1} = \\
&b_{u=u_1} \cdot V_{t,u=u_1,e=e_1}^* + b_{u=u_2} \cdot V_{t,u=u_2,e=e_1}^* + b_{u=u_3} \cdot V_{t,u=u_3,e=e_1}^* \}
\end{aligned} \tag{E.13}$$

are feasible in Equation (E.1). $\mathbf{V}_{t,e}^*$ is an optimal solution of a linear program on a facet of the feasible set. Accordingly, all points on that facet (points in M) are optimal in Equation (E.1). Note that again in the last equation of M the constant parts $a_{u=u_1}$, $a_{u=u_2}$ and $a_{u=u_3}$ are shortened. Following the definition of the Equation (E.8), we define the change in energy form e_1 until an energy conversion unit reaches its operational limits:

$$\begin{aligned}\Delta \mathrm{V}_{e=e_1}^{\text{Case4}} = \min\{&V^+_{t,u=u_1,e=e_1}(\mathbf{o}_t^*) - V^*_{t,u=u_1,e=e_1},\\ &\frac{b_{u=u_2}-b_{u=u_3}}{b_{u=u_2}\cdot(b_{u=u_1}-b_{u=u_3})}\cdot(V^*_{t,u=u_2,e=e_2} - V^-_{t,u=u_2,e=e_2}(\mathbf{o}_t^*)),\\ &\frac{b_{u=u_2}-b_{u=u_3}}{b_{u=u_3}\cdot(b_{u=u_1}-b_{u=u_2})}\cdot(V^+_{t,u=u_3,e=e_2}(\mathbf{o}_t^*) - V^*_{t,u=u_3,e=e_2})\}.\end{aligned} \tag{E.14}$$

With $\Delta \mathrm{V}_{e=e_1}^{\text{Case4}}$ we can manipulate the supplied energy by the energy conversion units such that the point $(\mathbf{o}_t^*, \tilde{\mathbf{V}}_{t,e})$ with $\tilde{V}_{t,u,e} = V^*_{t,u,e}, \forall u \in U \setminus \{u_1, u_2, u_3\}, e \in E_u$ and

$$\begin{aligned}\tilde{V}_{t,u=u_1,e=e_1} &= V^*_{t,u=u_1,e=e_1} + \Delta \mathrm{V}_{e=e_1}^{\text{Case4}}\\ \tilde{V}_{t,u=u_2,e=e_1} &= V^*_{t,u=u_2,e=e_1} - \frac{b_{u=u_1}-b_{u=u_3}}{b_{u=u_2}-b_{u=u_3}}\cdot\Delta \mathrm{V}_{e=e_1}^{\text{Case4}}\\ \tilde{V}_{t,u=u_3,e=e_1} &= V^*_{t,u=u_3,e=e_1} + \frac{b_{u=u_1}-b_{u=u_2}}{b_{u=u_2}-b_{u=u_3}}\cdot\Delta \mathrm{V}_{e=e_1}^{\text{Case4}}\\ \tilde{V}_{t,u=u_1,e=e_2} &= V^*_{t,u=u_1,e=e_2} + b_{u=u_1}\cdot\Delta \mathrm{V}_{e=e_1}^{\text{Case4}}\\ \tilde{V}_{t,u=u_2,e=e_2} &= V^*_{t,u=u_2,e=e_2} - \frac{b_{u=u_2}\cdot(b_{u=u_1}-b_{u=u_3})}{b_{u=u_2}-b_{u=u_3}}\cdot\Delta \mathrm{V}_{e=e_1}^{\text{Case4}}\\ \tilde{V}_{t,u=u_3,e=e_2} &= V^*_{t,u=u_3,e=e_2} + \frac{b_{u=u_3}\cdot(b_{u=u_1}-b_{u=u_2})}{b_{u=u_2}-b_{u=u_3}}\cdot\Delta \mathrm{V}_{e=e_1}^{\text{Case4}}\end{aligned} \tag{E.15}$$

lies within M and satisfies either $\tilde{V}_{t,u=u_1,e_1} = V^+_{t,u=u_1,e_1}(\mathbf{o}_t^*)$, $\tilde{V}_{t,u=u_2,e_2} = V^-_{t,u=u_2,e_2}(\mathbf{o}_t^*)$ or $\tilde{V}_{t,u=u_3,e_2} = V^+_{t,u=u_3,e_2}(\mathbf{o}_t^*)$, proving the desired property. With these changes in the supplied energy, only two energy conversion units are not operated at their operational limits.

Finally, if there are more than three energy conversion units $u \in U$ with $\mathrm{V}^-_{t,u,e}(\mathbf{o}_t^*) < V^*_{t,u,e} < \mathrm{V}^+_{t,u,e}(\mathbf{o}_t^*)$, the above constructions can be applied successively to three energy conversion units until the same result is reached. □

Appendix F

Publications and student theses

List of publications

Journal papers

Leenders, L., Hagedorn, D. F., Djelassi, H., Bardow, A., and Mitsos, A. (2022). Bilevel optimization for joint scheduling of production and energy systems. *Optimization & Engineering*.

Kämper, A., Holtwerth, A., Leenders, L., and Bardow, A. (2021a). AutoMoG 3D: Automated Data-Driven Model Generation of Multi-Energy Systems Using Hinging Hyperplanes. *Frontiers in Energy Research*, 9, 719658.

Leenders, L., Ganz, K., Bahl, B., Hennen, M., Baumgärtner, N., and Bardow, A. (2021). Scheduling coordination of multiple production and utility systems in a multi-leader multi-follower Stackelberg game. *Computers & Chemical Engineering*, 150, 107321.

Kämper, A., Leenders, L., Bahl, B., and Bardow, A. (2021b). AutoMoG: Automated data-driven Model Generation of multi-energy systems using piecewise-linear regression. *Computers & Chemical Engineering*, 145, 107162.

Dannapfel, M., Vierschilling, S. P., Leenders, L. and Saraßa, P. (2019). Integrierte Entscheidungen in der Fabrikplanung. *Zeitschrift für wirtschaftlichen Fabrikbetrieb*, 114(9), 521-524.

Leenders, L., Bahl, B., Lampe, M., Hennen, M., and Bardow, A. (2019b). Optimal design of integrated batch production and utility systems. *Computers & Chemical Engineering*, 128, 496-511.

Leenders, L., Bahl, B., Hennen, M., and Bardow, A. (2019a). Coordinating scheduling of production and utility system using a Stackelberg game. *Energy*, 175, 1283-1295.

Peer-reviewed conference papers

Nolzen, N., Ganter, A., Baumgärtner, N., Leenders, L., and Bardow, A. (accepted). Monetizing Flexibility in Day-Ahead and Continuous Intraday Electricity Markets. *14th International Symposium on Process Systems Engineering – PSE 2021+*.

Kämper, A., Geers, P., Leenders, L., and Bardow, A. (2021). Adaptive Rolling Horizon for operational optimization of multi-energy systems. In *ECOS 2021 - Proceedings of the 34th International Conference on Efficiency, Cost, Optimization, Simulation and Environmental Impact of Energy Systems*.

Nolzen, N., Leenders, L., and Bardow, A. (2021). Flexibility-expansion planning for enhanced balancing-power market participation of decentralized energy systems. In Türkay, M. and Aydın, E., editors *Proceedings of the 31th European Symposium on Computer Aided Process Engineering (ESCAPE-31)*, volume 50 of *Computer Aided Chemical Engineering*, pages 1841–1846.

Leenders, L., Starosta, A., Baumgärtner, N., and Bardow, A. (2020). Integrated scheduling of batch production and utility systems for provision of control reserve. In Yokoyama, R., Amano, Y., editors, *ECOS 2020 - Proceedings of the 33rd International Conference on Efficiency, Cost, Optimisation, Simulation and Environmental Impact of Energy Systems*, pages 712-723.

Hüttermann, A., Leenders, L., Bahl, B., and Bardow, A. (2019). Automated data-driven model generation of energy systems using piecewise linear regression. In Stanek, W., Gładysz, P., Werle, S., Adamczyk, W., editors, *ECOS 2019 - Proceedings of the 32nd International Conference on Efficiency, Cost, Optimization, Simulation and Environmental Impact of Energy Systems*, pages 327-338.

Leenders, L., Ganz, K., Bahl, B., Hennen, M., and Bardow, A. (2019). Coordination of multiple production and utility systems in a multi-leader multi-follower Stackelberg game. In Kiss, A. A., Zondervan, E., Lakerveld, R., Özkan, L., editors, *Proceedings of the 29th European Symposium on Computer Aided Process Engineering (ESCAPE-29)*, volume 46 of *Computer Aided Chemical Engineering*, pages 697-702.

Holters, L., Bahl, B., Hennen, M., and Bardow, A. (2018). Playing Stackelberg games for minimal cost for production and utilities. In *ECOS 2018 - Proceedings of the 31st International Conference on Efficiency, Cost, Optimization, Simulation and Environmental Impact of Energy Systems*.

Holters, L., Bahl, B., Lampe, M., Hennen, M., and Bardow, A. (2017). Integrated Synthesis of Batch Plants and Utility Systems. In Espuña, A., Graells, M., Puigjaner, L., editors, *Proceedings of the 27th European Symposium on Computer Aided Process Engineering (ESCAPE-27)*, volume 40 of *Computer Aided Chemical Engineering*, pages 625-630.

Further conference contributions

Nolzen, N., Ganter, A., Baumgärtner, N., Leenders, L., and Bardow, A. (accepted). Optimized market participation of flexible multi-energy systems in balancing-power, day-ahead, and continuous intraday markets. *International Conference on Operations Research - OR 2022.*

Mayer, P., Leenders, L., Shu, D., Winter, B., Zibunas, C., Reinert, C. and Bardow, A. (accepted). Transition paths towards a CO_2-based Chemical Industry within a Sector-Coupled Energy System. In Montastruc, L. and Negny, S., editors, *Book of Abstracts of the 32nd European Symposium on Computer Aided Process Engineering (ESCAPE-32).*

Shu, D. Y., Komesse, H. B., Beauchet, S., Leenders, L., Devaux, F., Santos-Moreau, V. and Bardow, A. (accepted). Insights from Life Cycle Assessment of the carbon capture and storage supply chain in the DMX™ Demonstration in Dunkirk (3D) project. *16th International Conference on Greenhouse Gas Control Technologies GHGT-16.*

Mayer, P., Leenders, L., Shu, D., Winter, B., Zibunas, C., Reinert, C., and Bardow, A. (2021). Transition Paths towards a CO_2-based Chemical Industry within a Sector-Coupled Energy System, *Climate Policy and Energy System Transformation: New Opportunities and Challenges of the Consideration of Co-Benefits*, Freiberg, Germany.

Nolzen, N., Leenders, L., and Bardow, A. (2020). Flexibility-expansion planning for decentralized energy systems to participate in balancing-power markets, *Computer Aided Process Engineering FORUM 2020 – CAPE 2020*, Kgs. Lyngby, Denmark.

Hüttermann, A., Leenders, L., Bahl, B., and Bardow, A. (2019). Automated data-driven model generation of energy systems using piecewise linear regression. *Jahrestreffen Forschungsnetzwerk Energiesystemanalyse*, Aachen, Germany.

Leenders, L., Bahl, B., Hennen, M., and Bardow, A. (2018). Gleichzeitige Minimierung von Produktions- und Energiekosten durch zeit- und lastabhängige Energiepreise, *Jahrestreffen der Fachgemeinschaft Prozess-, Apparate- und Anlagentechnik*, Cologne, Germany.

Hennen, M., Bahl, B., Hollermann, D. E., Holters, L., and Bardow, A. (2017). Optimal design of distributed energy supply systems, *Sustainable Process Integration Laboratory Scientific Conference (SPIL 2017)*, Brno, Czech Republic.

Student thesis supervised during this work

Sollich, M. (In preparation). Quantum-computing-based optimization of sustainable energy systems via decomposition. Master's thesis, ETH Zürich and RWTH Aachen University.

Casali, A. (2021). Techno-Economic Optimization of Hydrogen Utilization within a Sector-Coupled System in Nothern Netherlands. Semester project, ETH Zürich.

Lienhard, N. (2021). Modelling and Analysis of the Future Energy Supply in Switzerland within the European Electricity System. Master's thesis, ETH Zürich.

Hoffrogge, D. (2019). Gemischt-ganzzahlige Bilevel Optimierung von Produktionssystemen mit dezentraler Energieversorgung (in German). Master's thesis, RWTH Aachen University.

Starosta, A. (2019). Betriebsoptimierung von integrierten Produktions- und dezentralen Energiesystemen mit Unsicherheiten durch die Bereitstellung von Regelleistung (in German). Master's thesis, RWTH Aachen University.

Rauscher, S. (2018). Optimierter Stromhandel von Energie- und Produktionssystemen mit Demand-Side Management (in German). Master's thesis, RWTH Aachen University.

Ganz, K. (2018). Betriebsoptimierung mittels Multi-Leader Multi-Follower Spieltheorie von Produktionsstandorten mit dezentraler Energieversorgung (in German). Master's thesis, RWTH Aachen University.

Zilg, J. (2017). Entwicklung eines Modells zur Abbildung des lokalen Energiehandels und Integration in das Modell für virtuelle Kraftwerke (in German). Master's thesis, RWTH Aachen University.

Kohne, T. (2017). Mehrkriterielle Optimierung der operativen Planung von Produktionssystemen mit integriertem dezentralen Energieversorgungssystem (in German). Master's thesis, RWTH Aachen University.

Bibliography

50Hertz Transmission GmbH, Amprion GmbH, TenneT TSO GmbH, and TransnetBW GmbH (2019). Internetplattform zur Vergabe von Regelleistung. https://www.regelleistung.net. Access date: 25.7.2020.

Ackermann, S., Fumero, Y., and Montagna, J. M. (2021). New Problem Representation for the Simultaneous Resolution of Batching and Scheduling in Multiproduct Batch Plants. *Ind. Eng. Chem. Res.*, 60(6):2523–2535.

Agha, M. H., Thery, R., Hetreux, G., Hait, A., and Le Lann, J. M. (2010). Integrated production and utility system approach for optimizing industrial unit operations. *Energy*, 35(2):611–627.

Allman, A. and Zhang, Q. (2020). Distributed cooperative industrial demand response. *Journal of Process Control*, 86:81–93.

Alsalloum, H., Merghem-Boulahia, L., and Rahim, R. (2020). Hierarchical system model for the energy management in the smart grid: A game theoretic approach. *Sustainable Energy, Grids and Networks*, 21:100329.

Andiappan, V. (2017). State-Of-The-Art Review of Mathematical Optimisation Approaches for Synthesis of Energy Systems. *Process Integration and Optimization for Sustainability*, 1(3):165–188.

Atzeni, I., Ordonez, L. G., Scutari, G., Palomar, D. P., and Fonollosa, J. R. (2013). Demand-Side Management via Distributed Energy Generation and Storage Optimization. *IEEE Transactions on Smart Grid*, 4(2):866–876.

Avraamidou, S. and Pistikopoulos, E. N. (2019). A Bi-Level Formulation And Solution Method For The Integration Of Process Design And Scheduling. In Salvador Garcia Muñoz, Carl Laird, and Matthew Realff, editors, *Proceedings of the 9th International Conference on Foundations of Computer-Aided Process Design*, volume 47, pages 17–22. Elsevier.

Bahl, B., Lampe, M., Voll, P., and Bardow, A. (2017). Optimization-based identification and quantification of demand-side management potential for distributed energy supply systems. *Energy*, 135:889–899.

Bahl, B., Söhler, T., Hennen, M., and Bardow, A. (2018). Typical Periods for Two-Stage Synthesis by Time-Series Aggregation with Bounded Error in Objective Function. *Frontiers in Energy Research*, 5:1–13.

Bahrami, S. and Sheikhi, A. (2016). From Demand Response in Smart Grid Toward Integrated Demand Response in Smart Energy Hub. *IEEE Transactions on Smart Grid*, 7(2):650–658.

Barbosa-Póvoa, A. P. (2007). A critical review on the design and retrofit of batch plants. *Computers & Chemical Engineering*, 31(7):833–855.

Bard, J. F. (1991). Some properties of the bilevel programming problem. *Journal of Optimization Theory and Applications*, 68(2):371–378.

Basán, N. P., Grossmann, I. E., Gopalakrishnan, A., Lotero, I., and Méndez, C. A. (2018). Novel MILP Scheduling Model for Power-Intensive Processes under Time-Sensitive Electricity Prices. *Ind. Eng. Chem. Res.*, 57(5):1581–1592.

Baumgärtner, N., Delorme, R., Hennen, M., and Bardow, A. (2019a). Design of low-carbon utility systems: Exploiting time-dependent grid emissions for climate-friendly demand-side management. *Applied Energy*, 247:755–765.

Baumgärtner, N., Temme, F., Bahl, B., Hennen, M., Hollermann, D., and Bardow, A. (2019b). RiSES4 Rigorous Synthesis of Energy Supply Systems with Seasonal Storage by relaxation and time- series aggregation to typical periods. In *Proceedings of ECOS 2019 - The 32nd international conference on efficiency, cost, optimization, simulation and environmental impact of energy systems. June 23-28, 2019 Wroclaw, Poland*, Wroclaw, Poland.

Bertsimas, D. and Sim, M. (2004). The Price of Robustness. *Operations Research*, 52(1):35–53.

Birge, J. R. and Louveaux, F. (2011). *Introduction to Stochastic Programming*. Springer New York, New York, NY.

Bohlayer, M., Fleschutz, M., Braun, M., and Zöttl, G. (2020). Energy-intense production-inventory planning with participation in sequential energy markets. *Applied Energy*, 258:113954.

Boyd, S., Parikh, N., Chu, E., Peleato, B., and Eckstein, J. (2010). Distributed Optimization and Statistical Learning via the Alternating Direction Method of Multipliers. *Foundations and Trends® in Machine Learning*, 3(1):1–122.

Bundesnetzagentur | SMARD.de (2020). SMARD - Strommarktdaten. https://www.smard.de/. Access date: 26.10.2020.

Castro, P. M., Grossmann, I. E., and Zhang, Q. (2018). Expanding scope and computational challenges in process scheduling. *Computers & Chemical Engineering*, 114:14–42.

Castro, P. M., Harjunkoski, I., and Grossmann, I. E. (2011). Optimal scheduling of continuous plants with energy constraints. *Computers & Chemical Engineering*, 35(2):372–387.

Cheng, R., Forbes, J. F., and Yip, W. S. (2007). Price-driven coordination method for solving plant-wide MPC problems. *Journal of Process Control*, 17(5):429–438.

Demirhan, C. D., Tso, W. W., Ogumerem, G. S., and Pistikopoulos, E. N. (2019). Energy systems engineering - a guided tour. *BMC Chemical Engineering*, 1(1).

Dempe, S. and Zemkoho, A., editors (2020). *Bilevel Optimization*. Springer Optimization and Its Applications. Springer International Publishing, Cham.

Dias, L. S. and Ierapetritou, M. G. (2016). Integration of scheduling and control under uncertainties: Review and challenges. *Chemical Engineering Research and Design*, 116:98–113.

Dias, L. S. and Ierapetritou, M. G. (2017). From process control to supply chain management: An overview of integrated decision making strategies. *Computers & Chemical Engineering*, 106:826–835.

Djelassi, H., Glass, M., and Mitsos, A. (2019). Discretization-based algorithms for generalized semi-infinite and bilevel programs with coupling equality constraints. *Journal of Global Optimization*, 75(2):341–392.

Djelassi, H. and Mitsos, A. (2019). libALE - a library for algebraic-logical expression trees.

Drud, A. S. (1996). CONOPT: a system for large scale nonlinear optimisation. *Reference manual for CONPOT subroutine library*.

Engell, S., Paulen, R., Reniers, M. A., Sonntag, C., and Thompson, H. (2015). Core Research and Innovation Areas in Cyber-Physical Systems of Systems. In Mousavi, M. R. and Berger, C., editors, *Cyber Physical Systems. Design, Modeling, and Evaluation*, volume 9361, pages 40–55, Cham. Springer International Publishing.

Fischetti, M., Ljubić, I., Monaci, M., and Sinnl, M. (2016). Intersection Cuts for Bilevel Optimization. In Louveaux, Q. and Skutella, M., editors, *Integer Programming and Combinatorial Optimization*, volume 9682, pages 77–88. Springer International Publishing.

Floudas, C. A. and Lin, X. (2004). Continuous-time versus discrete-time approaches for scheduling of chemical processes: a review. *Computers & Chemical Engineering*, 28(11):2109–2129.

Frangopoulos, C. A. (2018). Recent developments and trends in optimization of energy systems. *Energy*, 164:1011–1020.

Frangopoulos, C. A., von Spakovsky, M. R., and Sciubba, E. (2002). A Brief Review of Methods for the Design and Synthesis Optimization of Energy Systems. *Int J Appl Thermodyn*, 5(4):151–160.

Frontier Economics (2016). METIS Technical Note T4: Overview of European Electricity Markets.

Gahm, C., Denz, F., Dirr, M., and Tuma, A. (2016). Energy-efficient scheduling in manufacturing companies: A review and research framework. *European Journal of Operational Research*, 248(3):744–757.

Gamrath, G. and Lübbecke, M. E. (2010). Experiments with a Generic Dantzig-Wolfe Decomposition for Integer Programs. In Festa, P., editor, *Experimental Algorithms: 9th international symposium, SEA 2010, Ischia Island, Naples, Italy, May 20-22, 2010 Proceedings*, volume 6049 of *LNCS sublibrary*, pages 239–252. Springer, Berlin and New York.

GAMS Development (2020). General Algebraic Modeling System (GAMS).

Gleixner, A., Bastubbe, M., Eifler, L., Gally, T., Gamrath, G., Gottwald, R. L., Hendel, G., Hojny, C., Koch, T., Lübbecke, M., Maher, S. J., Miltenberger, M., Müller, B., Pfetsch, M., Puchert, C., Rehfeldt, D., Schlösser, F., Schubert, C., Serrano, F., Shinano, Y., Viernickel, J. M., Wegscheider, F., Walter, M., Witt, J. T., and Witzig, J. (2018). The SCIP Optimization Suite 6.0.

Glover, F. (1975). Improved Linear Integer Programming Formulations of Nonlinear Integer Problems. *Management Science*, 22(4):455–460.

Goderbauer, S., Bahl, B., Voll, P., Lübbecke, M. E., Bardow, A., and Koster, A. M. (2016). An adaptive discretization MINLP algorithm for optimal synthesis of decentralized energy supply systems. *Computers & Chemical Engineering*, 95:38–48.

Grossmann, I. E., Apap, R. M., Calfa, B. A., García-Herreros, P., and Zhang, Q. (2016). Recent advances in mathematical programming techniques for the optimization of process systems under uncertainty. *Computers & Chemical Engineering*, 91:3–14.

Hadera, H., Ekström, J., Sand, G., Mäntysaari, J., Harjunkoski, I., and Engell, S. (2019). Integration of production scheduling and energy-cost optimization using Mean Value Cross Decomposition. *Computers & Chemical Engineering*, 129:106436.

Hadera, H., Harjunkoski, I., Sand, G., Grossmann, I. E., and Engell, S. (2015). Optimization of steel production scheduling with complex time-sensitive electricity cost. *Computers & Chemical Engineering*, 76:117–136.

Hadera, H., Labrik, R., Mäntysaari, J., Sand, G., Harjunkoski, I., and Engell, S. (2016). Integration of Energy-cost Optimization and Production Scheduling Using Multiparametric Programming. In *26th European Symposium on Computer Aided Process Engineering*, volume 38 of *Computer Aided Chemical Engineering*, pages 559–564. Elsevier.

Hall, R. (2012). *Handbook of Healthcare System Scheduling*, volume 168. Springer US, Boston, MA.

Harjunkoski, I., Maravelias, C. T., Bongers, P., Castro, P. M., Engell, S., Grossmann, I. E., Hooker, J., Méndez, C., Sand, G., and Wassick, J. (2014). Scope for industrial applications of production scheduling models and solution methods. *Computers & Chemical Engineering*, 62:161–193.

Hemmati, M. and Smith, J. C. (2016). A mixed-integer bilevel programming approach for a competitive prioritized set covering problem. *Discrete Optimization*, 20:105–134.

Higle, J. L. (2005). Stochastic Programming: Optimization When Uncertainty Matters. In Greenberg, H. J. and Smith, J. C., editors, *Emerging Theory, Methods, and Applications*, pages 30–53. INFORMS.

Holmberg, K. (1992). Linear mean value cross decomposition: A generalization of the Kornai-Liptak method. *European Journal of Operational Research*, 62(1):55–73.

Hubbs, C. D., Li, C., Sahinidis, N. V., Grossmann, I. E., and Wassick, J. M. (2020). A deep reinforcement learning approach for chemical production scheduling. *Computers & Chemical Engineering*, 141:106982.

IBM Corporation (2020). IBM ILOG CPLEX Optimization Studio.

IPCC (2014). Climate change 2014: Mitigation of climate change Working Group III contribution to the Fifth Assessment Report of the Intergovernmental Panel on Climate Change.

Janak, S. L., Floudas, C. A., Kallrath, J., and Vormbrock, N. (2006). Production Scheduling of a Large-Scale Industrial Batch Plant. II. Reactive Scheduling. *Ind. Eng. Chem. Res.*, 45(25):8253–8269.

Jeroslow, R. G. (1985). The polynomial hierarchy and a simple model for competitive analysis. *Mathematical Programming*, 32(2):146–164.

Jose, R. A. and Ungar, L. H. (2000). Pricing interprocess streams using slack auctions. *AIChE J*, 46(3):575–587.

Kallrath, J. (2002). Planning and scheduling in the process industry. *OR Spectr*, 24(3):219–250.

Kämper, A., Holtwerth, A., Leenders, L., and Bardow, A. (2021a). AutoMoG 3D: Automated Data-Driven Model Generation of Multi-Energy Systems Using Hinging Hyperplanes. *Frontiers in Energy Research*, 9.

Kämper, A., Leenders, L., Bahl, B., and Bardow, A. (2021b). AutoMoG: Automated data-driven Model Generation of multi-energy systems using piecewise-linear regression. *Computers & Chemical Engineering*, 145:107162.

Klein Haneveld, W. K., van der Vlerk, M. H., and Romeijnders, W., editors (2020). *Stochastic Programming*. Graduate Texts in Operations Research. Springer International Publishing, Cham.

Klemeš, J. J. and Kravanja, Z. (2013). Forty years of Heat Integration: Pinch Analysis (PA) and Mathematical Programming (MP). *Current Opinion in Chemical Engineering*, 2(4):461–474.

Kleniati, P.-M. and Adjiman, C. S. (2014a). Branch-and-Sandwich: a deterministic global optimization algorithm for optimistic bilevel programming problems. Part I: Theoretical development. *Journal of Global Optimization*, 60(3):425–458.

Kleniati, P.-M. and Adjiman, C. S. (2014b). Branch-and-Sandwich: a deterministic global optimization algorithm for optimistic bilevel programming problems. Part II: Convergence analysis and numerical results. *Journal of Global Optimization*, 60(3):459–481.

Kleniati, P.-M. and Adjiman, C. S. (2015). A generalization of the Branch-and-Sandwich algorithm: From continuous to mixed-integer nonlinear bilevel problems. *Computers & Chemical Engineering*, 72:373–386.

Kocis, G. R. and Grossmann, I. E. (1989). Computational experience with dicopt solving MINLP problems in process systems engineering. *Computers & Chemical Engineering*, 13(3):307–315.

Kondili, E., Pantelides, C. C., and Sargent, R. (1993). A general algorithm for short-term scheduling of batch operations—I. MILP formulation. *Computers & Chemical Engineering*, 17(2):211–227.

Kopanos, G. M. and Puigjaner, L. (2019). *Solving Large-Scale Production Scheduling and Planning in the Process Industries*. Springer International Publishing, Cham.

Kostarelou, E. and Kozanidis, G. (2021). Bilevel programming solution algorithms for optimal price-bidding of energy producers in multi-period day-ahead electricity markets with non-convexities. *Optimization and Engineering*, 22(1):449–484.

Kotzur, L., Nolting, L., Hoffmann, M., Groß, T., Smolenko, A., Priesmann, J., Büsing, H., Beer, R., Kullmann, F., Singh, B., Praktiknjo, A., Stolten, D., and Robinius, M. (2020). A modeler's guide to handle complexity in energy system optimization.

Kumbartzky, N., Schacht, M., Schulz, K., and Werners, B. (2017). Optimal operation of a CHP plant participating in the German electricity balancing and day-ahead spot market. *European Journal of Operational Research*, 261(1):390–404.

Lee, H. and Maravelias, C. T. (2017). Discrete-time mixed-integer programming models for short-term scheduling in multipurpose environments. *Computers & Chemical Engineering*, 107:171–183.

Leenders, L., Bahl, B., Hennen, M., and Bardow, A. (2019a). Coordinating scheduling of production and utility system using a Stackelberg game. *Energy*, 175:1283–1295.

Leenders, L., Bahl, B., Lampe, M., Hennen, M., and Bardow, A. (2019b). Optimal design of integrated batch production and utility systems. *Computers & Chemical Engineering*, 128:496–511.

Leenders, L., Ganz, K., Bahl, B., Hennen, M., Baumgärtner, N., and Bardow, A. (2021). Scheduling coordination of multiple production and utility systems in a multi-leader multi-follower Stackelberg game. *Computers & Chemical Engineering*, 150:107321.

Leenders, L., Hagedorn, D. F., Djelassi, H., Bardow, A., and Mitsos, A. (2022). Bilevel optimization for joint scheduling of production and energy systems. *Optimization and Engineering*.

Leenders, L., Starosta, A., Baumgärtner, N., and Bardow, A. (2020). Integrated scheduling of batch production and utility systems for provision of control reserve. In *Proceedings of ECOS 2020 - 33rd International Conference on Efficiency, Cost, Optimization, Simulation and Environmental Impact of Energy Systems*, pages 712–723. Osaka, Japan.

Leo, E., Dalle Ave, G., Harjunkoski, I., and Engell, S. (2021). Stochastic short-term integrated electricity procurement and production scheduling for a large consumer. *Computers & Chemical Engineering*, 145:107191.

Letsios, D., Baltean-Lugojan, R., Ceccon, F., Mistry, M., Wiebe, J., and Misener, R. (2020). Approximation algorithms for process systems engineering. *Computers & Chemical Engineering*, 132:106599.

Li, H., Lü, Q., Wang, Z., Liao, X., and Huang, T. (2020). *Distributed Optimization: Advances in Theories, Methods, and Applications*. Springer Singapore, Singapore.

Liew, P. Y., Theo, W. L., Wan Alwi, S. R., Lim, J. S., Abdul Manan, Z., Klemeš, J. J., and Varbanov, P. S. (2017). Total Site Heat Integration planning and design for industrial, urban and renewable systems. *Renewable and Sustainable Energy Reviews*, 68:964–985.

Magege, S. R. and Majozi, T. (2021). A comprehensive framework for synthesis and design of heat-integrated batch plants: Consideration of intermittently-available streams. *Renewable and Sustainable Energy Reviews*, 135:110125.

Maharjan, S., Zhu, Q., Zhang, Y., Gjessing, S., and Basar, T. (2013). Dependable Demand Response Management in the Smart Grid: A Stackelberg Game Approach. *IEEE Transactions on Smart Grid*, 4(1):120–132.

Maharjan, S., Zhu, Q., Zhang, Y., Gjessing, S., and Basar, T. (2016). Demand Response Management in the Smart Grid in a Large Population Regime. *IEEE Transactions on Smart Grid*, 7(1):189–199.

Majewski, D. E., Lampe, M., Voll, P., and Bardow, A. (2017a). TRusT: A Two-stage Robustness Trade-off approach for the design of decentralized energy supply systems. *Energy*, 118:590–599.

Majewski, D. E., Wirtz, M., Lampe, M., and Bardow, A. (2017b). Robust multi-objective optimization for sustainable design of distributed energy supply systems. *Computers & Chemical Engineering*, 102:26–39.

Majozi, T., Seid, E. R., and Lee, J.-Y. (2015). *Synthesis, Design, and Resource Optimization in Batch Chemical Plants*. CRC Press.

Mancarella, P. (2014). MES (multi-energy systems): An overview of concepts and evaluation models. *Energy*, 65:1–17.

Maravelias, C. T. (2012). General framework and modeling approach classification for chemical production scheduling. *AIChE Journal*, 58(6):1812–1828.

Mavromatidis, G., Orehounig, K., and Carmeliet, J. (2018). Design of distributed energy systems under uncertainty: A two-stage stochastic programming approach. *Applied Energy*, 222:932–950.

McKay, M. D., Beckman, R. J., and Conover, W. J. (2000). A Comparison of Three Methods for Selecting Values of Input Variables in the Analysis of Output from a Computer Code. *Technometrics*, 42(1):55–61.

Méndez, C. A., Cerdá, J., Grossmann, I. E., Harjunkoski, I., and Fahl, M. (2006). State-of-the-art review of optimization methods for short-term scheduling of batch processes. *Computers & Chemical Engineering*, 30(6-7):913–946.

Merkert, L., Harjunkoski, I., Isaksson, A., Säynevirta, S., Saarela, A., and Sand, G. (2015). Scheduling and energy – Industrial challenges and opportunities. *Computers & Chemical Engineering*, 72:183–198.

Mitra, S., Grossmann, I. E., Pinto, J. M., and Arora, N. (2012). Optimal production planning under time-sensitive electricity prices for continuous power-intensive processes. *Computers & Chemical Engineering*, 38:171–184.

Mitsos, A. (2010). Global solution of nonlinear mixed-integer bilevel programs. *Journal of Global Optimization*, 47(4):557–582.

Mitsos, A., Asprion, N., Floudas, C. A., Bortz, M., Baldea, M., Bonvin, D., Caspari, A., and Schäfer, P. (2018). Challenges in process optimization for new feedstocks and energy sources. *Computers & Chemical Engineering*, 113:209–221.

Mitsos, A., Lemonidis, P., and Barton, P. I. (2008). Global solution of bilevel programs with a nonconvex inner program. *Journal of Global Optimization*, 42(4):475–513.

Mondal, A., Misra, S., Patel, L. S., Pal, S. K., and Obaidat, M. S. (2018). DEMANDS: Distributed Energy Management Using Noncooperative Scheduling in Smart Grid. *IEEE Systems Journal*, 12(3):2645–2653.

Motalleb, M., Annaswamy, A., and Ghorbani, R. (2018). A real-time demand response market through a repeated incomplete-information game. *Energy*, 143:424–438.

Muche, T., Höge, C., Renner, O., and Pohl, R. (2016). Profitability of participation in control reserve market for biomass-fueled combined heat and power plants. *Renewable Energy*, 90:62–76.

Nolzen, N., Leenders, L., and Bardow, A. (2021). Flexibility-expansion planning for enhanced balancing-power market participation of decentralized energy systems. In Türkay, M. and Aydın, E., editors, *Proceedings of the 31st European Symposium on Computer Aided Process Engineering*, volume 50, pages 1841–1846.

Osborne, M. J. and Rubinstein, A. (1994). *A course in game theory*. MIT Press, Cambridge, Mass. and London.

Pablos, C., Merino, A., Acebes, L. F., Pitarch, J. L., and Biegler, L. T. (2021). Dynamic optimization approach to coordinate industrial production and cogeneration operation under electricity price fluctuations. *Computers & Chemical Engineering*, 149:107292.

Palensky, P. and Dietrich, D. (2011). Demand Side Management: Demand Response, Intelligent Energy Systems, and Smart Loads. *IEEE Transactions on Industrial Informatics*, 7(3):381–388.

Pantelides, C. C. (1994). Unified frameworks for optimal process planning and scheduling. In *Proceedings on the second conference on foundations of computer aided operations*, pages 253–274.

Papadopoulos, A. I., Tsivintzelis, I., Linke, P., and Seferlis, P. (2018). Computer-Aided Molecular Design: Fundamentals, Methods, and Applications. In *Reference Module in Chemistry, Molecular Sciences and Chemical Engineering*. Elsevier.

Papoulias, S. A. and Grossmann, I. E. (1983a). A structural optimization approach in process synthesis—I. *Computers & Chemical Engineering*, 7(6):695–706.

Papoulias, S. A. and Grossmann, I. E. (1983b). A structural optimization approach in process synthesis—II: Heat recovery networks. *Computers & Chemical Engineering*, 7(6):707–721.

Papoulias, S. A. and Grossmann, I. E. (1983c). A structural optimization approach in process synthesis—III: Total processing systems. *Computers & Chemical Engineering*, 7(6):723–734.

Pinedo, M. L. (2016). *Scheduling*. Springer International Publishing, Cham.

Pinto, T. (2003). Optimal design of heat-integrated multipurpose batch facilities with economic savings in utilities: a mixed integer mathematical formulation. *Ann. Oper. Res.*, 120:201–230.

Pinto, T., Barbósa-Póvoa, A. P. F. D., and Novais, A. Q. (2008). Design of Multipurpose Batch Plants: A Comparative Analysis between the STN, m-STN, and RTN Representations and Formulations. *Ind. Eng. Chem. Res.*, 47(16):6025–6044.

Pistikopoulos, E. N., Barbosa-Povoa, A., Lee, J. H., Misener, R., Mitsos, A., Reklaitis, G. V., Venkatasubramanian, V., You, F., and Gani, R. (2021). Process systems engineering – The generation next? *Computers & Chemical Engineering*, 147:107252.

Ramos, M. A., Boix, M., Aussel, D., Montastruc, L., and Domenech, S. (2016). Water integration in eco-industrial parks using a multi-leader-follower approach. *Computers & Chemical Engineering*, 87:190–207.

Ramos, M. A., Rocafull, M., Boix, M., Aussel, D., Montastruc, L., and Domenech, S. (2018). Utility network optimization in eco-industrial parks by a multi-leader follower game methodology. *Computers & Chemical Engineering*, 112:132–153.

Schäfer, P., Westerholt, H. G., Schweidtmann, A. M., Ilieva, S., and Mitsos, A. (2019). Model-based bidding strategies on the primary balancing market for energy-intense processes. *Computers & Chemical Engineering*, 120:4–14.

Seid, E. R. and Majozi, T. (2013). Design and Synthesis of Multipurpose Batch Plants Using a Robust Scheduling Platform. *Ind. Eng. Chem. Res.*, 52(46):16301–16313.

Seid, E. R. and Majozi, T. (2015). Design and synthesis of batch processing plants: A consideration of utility aspect and using a robust scheduling platform. *Computer Aided Chemical Engineering*, 37:1169–1174.

Shah, N., Pantelides, C. C., and Sargent, R. (1993). A general algorithm for short-term scheduling of batch operations—II. Computational issues. *Computers & Chemical Engineering*, 17(2):229–244.

Sobri, S., Koohi-Kamali, S., and Rahim, N. A. (2018). Solar photovoltaic generation forecasting methods: A review. *Energy Conversion and Management*, 156:459–497.

Soliman, H. M. and Leon-Garcia, A. (2014). Game-Theoretic Demand-Side Management With Storage Devices for the Future Smart Grid. *IEEE Transactions on Smart Grid*, 5(3):1475–1485.

Soyster, A. L. (1973). Technical Note—Convex Programming with Set-Inclusive Constraints and Applications to Inexact Linear Programming. *Operations Research*, 21(5):1154–1157.

Teichgraeber, H. and Brandt, A. R. (2019). Clustering methods to find representative periods for the optimization of energy systems: An initial framework and comparison. *Applied Energy*, 239:1283–1293.

Théry, R., Hétreux, G., Agha, M. H., Haït, A., and Le Lann, J. M. (2012). The extended resource task network: a framework for the combined scheduling of batch processes and CHP plants. *International Journal of Production Research*, 50(3):623–646.

Tsoukalas, A., Rustem, B., and Pistikopoulos, E. N. (2009a). A global optimization algorithm for generalized semi-infinite, continuous minimax with coupled constraints and bi-level problems. *Journal of Global Optimization*, 44(2):235–250.

Tsoukalas, A., Wiesemann, W., and Rustem, B. (2009b). Global optimisation of pessimistic bi-level problems. In Pardalos, P. and Coleman, T., editors, *Lectures on Global Optimization*, pages 215–243. American Mathematical Society, Providence, Rhode Island.

van Roy, T. J. (1983). Cross decomposition for mixed integer programming. *Mathematical Programming*, 25(1):46–63.

Voll, P., Klaffke, C., Hennen, M., and Bardow, A. (2013). Automated superstructure-based synthesis and optimization of distributed energy supply systems. *Energy*, 50:374–388.

von Stackelberg, H. (1934). *Marktform und Gleichgewicht*. Springer, Berlin.

Voss, S. and Woodruff, D. L. (2006). *Introduction to computational optimization models for production planning in a supply chain*. Springer, Berlin and New York, 2nd ed. edition.

Wang, H., Li, H., and Zhou, B. (2021). *Distributed Optimization, Game and Learning Algorithms*. Springer Singapore, Singapore.

Wang, Z., Gao, F., Zhai, Q., Guan, X., Liu, K., and Zhou, D. (2012). An integrated optimization model for generation and batch production load scheduling in energy intensive enterprise. In *2012 IEEE Power and Energy Society General Meeting*, pages 1–8, Piscataway, N.J. IEEE.

Wassick, J. M. (2009). Enterprise-wide optimization in an integrated chemical complex. *Computers & Chemical Engineering*, 33(12):1950–1963.

Wei, F., Jing, Z. X., Wu, P. Z., and Wu, Q. H. (2017). A Stackelberg game approach for multiple energies trading in integrated energy systems. *Applied Energy*, 200:315–329.

Wenzel, S., Paulen, R., Stojanovski, G., Krämer, S., Beisheim, B., and Engell, S. (2016). Optimal resource allocation in industrial complexes by distributed optimization and dynamic pricing. *at - Automatisierungstechnik*, 64(6).

Wenzel, S., Riedl, F., and Engell, S. (2020). An efficient hierarchical market-like coordination algorithm for coupled production systems based on quadratic approximation. *Computers & Chemical Engineering*, 134:106704.

Wiesemann, W., Tsoukalas, A., Kleniati, P.-M., and Rustem, B. (2013). Pessimistic Bilevel Optimization. *SIAM Journal on Optimization*, 23(1):353–380.

Wogrin, S., Pineda, S., and Tejada-Arango, D. A. (2020). Applications of Bilevel Optimization in Energy and Electricity Markets. In Dempe, S. and Zemkoho, A., editors, *Bilevel Optimization*, volume 161 of *Springer Optimization and Its Applications*, pages 139–168. Springer International Publishing, Cham.

Yeh, K., Whittaker, C., Realff, M. J., and Lee, J. H. (2015). Two stage stochastic bilevel programming model of a pre-established timberlands supply chain with biorefinery investment interests. *Computers & Chemical Engineering*, 73:141–153.

Yokoyama, R., Haizuka, K., and Wakui, T. (2019). Analysis on cooperation between central power utility and distributed cogeneration systems by bilevel mixed-integer linear programming. In *Proceedings of ECOS 2019 - The 32nd International Conference on Efficiency, Cost, Optimization, Simulation and Environmental Impact of Energy Systems*, pages 1675–1686. Wroclaw, Poland.

Yu, M. and Hong, S. H. (2016). Supply–demand balancing for power management in smart grid: A Stackelberg game approach. *Applied Energy*, 164:702–710.

Yue, D., Gao, J., Zeng, B., and You, F. (2019). A projection-based reformulation and decomposition algorithm for global optimization of a class of mixed integer bilevel linear programs. *Journal of Global Optimization*, 73(1):27–57.

Yue, D. and You, F. (2017). Stackelberg-game-based modeling and optimization for supply chain design and operations: A mixed integer bilevel programming framework. *Computers & Chemical Engineering*, 102:81–95.

Zeng, B. and An, Y. (2014). Solving Bilevel Mixed Integer Program by Reformulations and Decomposition.

Zhan, Z.-H., Liu, X.-F., Gong, Y.-J., Zhang, J., Chung, H. S.-H., and Li, Y. (2015). Cloud Computing Resource Scheduling and a Survey of Its Evolutionary Approaches. *ACM Comput Surv*, 47(4):1–33.

Zhang, Q., Cremer, J. L., Grossmann, I. E., Sundaramoorthy, A., and Pinto, J. M. (2016a). Risk-based integrated production scheduling and electricity procurement for continuous power-intensive processes. *Computers & Chemical Engineering*, 86:90–105.

Zhang, Q. and Grossmann, I. E. (2016). Enterprise-wide optimization for industrial demand side management: Fundamentals, advances, and perspectives. *Chemical Engineering Research and Design*, 116:114–131.

Zhang, Q., Grossmann, I. E., Heuberger, C. F., Sundaramoorthy, A., and Pinto, J. M. (2015). Air separation with cryogenic energy storage: Optimal scheduling considering electric energy and reserve markets. *AIChE Journal*, 61(5):1547–1558.

Zhang, Q., Martín, M., and Grossmann, I. E. (2019). Integrated design and operation of renewables-based fuels and power production networks. *Computers & Chemical Engineering*, 122:80–92.

Zhang, Q., Morari, M. F., Grossmann, I. E., Sundaramoorthy, A., and Pinto, J. M. (2016b). An adjustable robust optimization approach to scheduling of continuous industrial processes providing interruptible load. *Computers & Chemical Engineering*, 86:106–119.

Zhang, Q., Sundaramoorthy, A., Grossmann, I. E., and Pinto, J. M. (2016c). A discrete-time scheduling model for continuous power-intensive process networks with various power contracts. *Computers & Chemical Engineering*, 84:382–393.

Zhang, Y., Liu, F., Wang, Z., Su, Y., Wang, W., and Feng, S. (2021). Robust Scheduling of Virtual Power Plant under Exogenous and Endogenous Uncertainties.

Zhang, Y., Wang, J., and Wang, X. (2014). Review on probabilistic forecasting of wind power generation. *Renewable and Sustainable Energy Reviews*, 32:255–270.

Zhao, H., Rong, G., and Feng, Y. (2014). Multiperiod Planning Model for Integrated Optimization of a Refinery Production and Utility System. *Ind. Eng. Chem. Res.*, 53(41):16107–16122.

Zhao, S., Grossmann, I. E., and Tang, L. (2018). Integrated scheduling of rolling sector in steel production with consideration of energy consumption under time-of-use electricity prices. *Computers & Chemical Engineering*, 111:55–65.

Zhou, Z., Zhang, J., Liu, P., Li, Z., Georgiadis, M. C., and Pistikopoulos, E. N. (2013). A two-stage stochastic programming model for the optimal design of distributed energy systems. *Applied Energy*, 103(0):135–144.

Zulkafli, N. I. and Kopanos, G. M. (2016). Planning of production and utility systems under unit performance degradation and alternative resource-constrained cleaning policies. *Applied Energy*, 183:577–602.

Zulkafli, N. I. and Kopanos, G. M. (2017). Integrated condition-based planning of production and utility systems under uncertainty. *Journal of Cleaner Production*, 167:776–805.

Aachener Beiträge zur Technischen Thermodynamik

ABTT 1
Philip Voll
Automated Optimization-Based Synthesis of Distributed Energy Supply Systems
1. Auflage 2014
ISBN 978-3-86130-474-6

ABTT 2
Johannes Jung
Comparative Life Cycle Assessment of Industrial Multi-Product Processes
1. Auflage 2014
ISBN 978-3-86130-471-5

ABTT 3
Franz Lanzerath
Modellgestützte Entwicklung von Adsorptionswärmepumpen
1. Auflage 2014
ISBN 978-3-86130-472-2

ABTT 4
Thorsten Brands
Einfluss der Gemischzusammensetzung auf die Verbrennung im Diesel- und GCAI-Motor
1. Auflage 2014
ISBN 978-3-95886-006-3

ABTT 5
Dominique Dechambre
Efficient Measurement of Liquid-Liquid Equilibria using Automation and Optimal Experimental Design
1. Auflage 2016
ISBN 978-395886-077-3

ABTT 6
Niklas von der Aßen
From Life-Cycle Assesement towards life-Cycle Design of Carbon Dioxide Capture and Utilization
1. Auflage 2016
ISBN 978-3-95886-080-3

ABTT 7
Matthias Lampe
Integrated Process and Organic Rankine Cycle Working Fluid Design in the Continuous-Molecular Targeting Framework
1. Auflage 2016
ISBN 978-3-95886-086-5

ABTT 8
Thomas Hülser
Optische Untersuchung der Zündvorgänge und deren Auswirkung auf die Verbrennung in PKW-Motoren
1. Auflage 2016
ISBN 978-3-95886-090-2

Aachener Beiträge zur Technischen Thermodynamik

ABTT 9
Malte Döntgen
Reaction Models from Reactive Molecular Dynamics and High-Level Kinetics Predictions
1. Auflage 2016
ISBN 978-3-95886-156-5

ABTT 10
Heike Schreiber
Experiments and Validated Models for Adsorption Thermal Energy Storage in Industrial and Residential Application
1. Auflage 2017
ISBN 978-3-95886-178-7

ABTT 11
André Dirk Sternberg
System-Wide Perspective for Life Cycle Assesment of CO_2-based C1-Chemicals
1. Auflage 2017
ISBN 978-3-95886-193-0

ABTT 12
Uwe Bau
From Dynamic Simulation to Optimal Design and Control of Adsorption Energy Systems
1. Auflage 2018
ISBN 978-3-95886-216-6

ABTT 13
Christian Jens
Modellbasiertes Design von Produkt, Lösungsmittel und Prozess für die Ameisensäure-synthese aus CO_2 und H_2
1. Auflage 2018
ISBN 978-3-95886-231-9

ABTT 14
Jan David Scheffczyk
Integrated Computer-Aided Design of Molecules and Processes using COSMO-RS
1. Auflage 2018
ISBN 978-3-95886-236-4

ABTT 15
Björn Bahl
Optimization-Based Synthesis of Large-Scale Energy Systems by Time-Series Aggregation
1. Auflage 2018
ISBN 978-3-95886-240-1

Aachener Beiträge zur Technischen Thermodynamik

ABTT 16
Bastian Liebergesell
A Milliliter-Scale Setup for the Efficient Characterization of Multicomponent Vapor-Liquid Equilibria Using Raman Spectroscopy
1. Auflage 2018
ISBN 978-3-95886-247-0

ABTT 17
Stefan Wilhelm Graf
A Design Approach for Adsorption Energy Systems Integrating Dynamic Modeling with Small-Scale Experiments
1. Auflage 2018
ISBN 978-3-95886-258-6

ABTT 18
Sebastian Kaminski
Quantum-Mechanics-Based Prediction of SAFT Parameters for Non-Associating and Associating Molecules Containing Carbon, Hydrogen, Oxygen and Nitrogen
1. Auflage 2019
ISBN 978-3-95886-270-8

ABTT 19
Maike Renate Hennen
Decision Support for the Synthesis of Energy Systems by Analysis of the Near-Optimal Solution Space
1. Auflage 2019
ISBN 978-3-95886-277-7

ABTT 20
Peyman Yamin
COSMO-RS-Based Methods for Improved Modelling of Complex Chemical Systems
1. Auflage 2019
ISBN 978-3-95886-288-3

ABTT 21
Meltem Erdogan
Assessement of Adsorbents for Drying by Experiments and Dynamic Simulations
1. Auflage 2019
ISBN 978-3-95886-303-3

ABTT 22
Christian Schulz
SRS/LIF-Messungen zur Charakterisierung rußarmer dieselähnlicher Flammen von alternativen Kraftstoffen und n-Heptan
1. Auflage 2019
ISBN 978-3-91886-310-1

Aachener Beiträge zur Technischen Thermodynamik

ABTT 23
Peter Beumers
Physically-Based Models for the Analysis of Raman Spectra
1. Auflage 2019
ISBN 978-3-95886-319-4

ABTT 24
Arne Kätelhön
Technology Choice Model for Consequential Life Cycle Assessment
1. Auflage 2019
ISBN 978-3-95886-324-8

ABTT 25
Christine Peters
Measurement of Multicomponent Diffusion in Liquids Using Raman Microspectroscopy and Microfluidics
1. Auflage 2020
ISBN 978-3-95886-337-8

ABTT 26
Dinah Elena Hollermann
Reliable and Robust Optimal Design of Sustainable Energy Systems
1. Auflage 2020
ISBN 978-3-95886-346-0

ABTT 27
Thomas Raffius
Laserspektroskopische Analyse von selbstzündenden motorischen Einspritzstrahlen alternativer Biokraftstoffe
1. Auflage 2020
ISBN 978-3-95886-358-3

ABTT 28
Johannes Schilling
Integrated Thermo-Economic Design of Processes and Molecules Using PC-SAFT
1. Auflage 2020
ISBN 978-3-95886-368-2

ABTT 29
Nils Julius Baumgärtner
Optimization of Low-Carbon Energy Systems from Industrial to National Scale
1. Auflage 2020
ISBN 978-3-95886-385-9

ABTT 30
Ludger Wolff
From Model-based Experimental Design and Analysis of Diffusion and Liquid-Liquid Equilibria to Process Applications
1. Auflage 2020
ISBN 978-3-95886-402-3

Aachener Beiträge zur Technischen Thermodynamik

ABTT 31
Andrej Gibelhaus
A Model-based Framework for Optimal Systems Integration of Adsorption Chillers
1. Auflage 2021
ISBN 978-3-95886-406-1

ABTT 32
Jan Seiler
Debottlenecking the Evaporator in Water-Based Adsorption Chillers
1. Auflage 2021
ISBN 978-3-95886-407-8

ABTT 33
Leif Kröger
Prediction of Reaction Rate Constants for the Synthesis of Microgels
1. Auflage 2021
ISBN 978-395886-425-2

ABTT 34
Sarah von Pfingsten
Uncertainty Analysis in Matrix-Based Life Cycle Assessment
1. Auflage 2022
ISBN 978-3-95886-431-3

ABTT 35
Ludger Leenders
Optimization Methods for Integrating Energy and Production Systems
1. Auflage 2022
ISBN 978-3-95886-445-0